Carla Eberhardt Cerutti
Claudia Scherer Kuhn

Profitability of Agricultural Production

Carla Eberhardt Cerutti
Claudia Scherer Kuhn

Profitability of Agricultural Production

Analysis of the profitability of agricultural production on a small rural property

ScienciaScripts

Imprint

Any brand names and product names mentioned in this book are subject to trademark, brand or patent protection and are trademarks or registered trademarks of their respective holders. The use of brand names, product names, common names, trade names, product descriptions etc. even without a particular marking in this work is in no way to be construed to mean that such names may be regarded as unrestricted in respect of trademark and brand protection legislation and could thus be used by anyone.

Cover image: www.ingimage.com

This book is a translation from the original published under ISBN 978-613-9-61388-5.

Publisher:
Sciencia Scripts
is a trademark of
Dodo Books Indian Ocean Ltd. and OmniScriptum S.R.L publishing group

120 High Road, East Finchley, London, N2 9ED, United Kingdom
Str. Armeneasca 28/1, office 1, Chisinau MD-2012, Republic of Moldova, Europe
Printed at: see last page
ISBN: 978-620-7-65651-6

Table of contents:

DEDICATORY

I dedicate this Final Paper to all the people who have always been with me on this journey, encouraging me and supporting me so that I could complete another achievement in my life.

I dedicate it to my parents, Edgar Eberhardt and Odete Sigolini Eberhardt, for my life, for the motivation and encouragement to start my degree and for the opportunity to complete it, for being present at every moment of my life. I dedicate it to my boyfriend Joâo Alexandre Cerutti for transforming my life, giving me meaning and the strength to always pursue my goals.

ACKNOWLEDGMENTS

First of all, I thank God for the gift of life, for having important people in my life, for everything I have, for everything I've achieved.

I would like to thank my parents, my sister and my boyfriend for their companionship and encouragement in the pursuit of my goals.

To the teachers who helped me on this journey, for passing on knowledge and giving new meaning to my life.

To Professor Claudia Kuhn, who went out of her way to help me and assist me in the preparation of the course conclusion, for the knowledge she passed on during this period.

And to my colleagues who have been part of my life over the last four years, for their companionship and the friendships I have gained.

"The really good thing is to go into battle with determination, embrace life with passion, lose with class and win with daring. Because the world belongs to those who dare, and life is 'too much' to be insignificant."
Augusto Branco

SUMMARY

With the changes in technology, companies need to be constantly updating the technologies involved in production and management. In the same way, rural producers need to be up-to-date with the technology of the machines and the ways of managing the property, so that they are aware of the costs of their production, as well as their income and the returns on their property. This study seeks to identify the production process that brings the most return to the property, and whether current production is capable of generating the minimum return expected from each production. This study shows the results of production in relation to the minimum return on land rent in each period of the production process. The costs of the production processes were identified, as well as the profitability of each production and a comparison of the production alternatives that bring the highest return to the property. To carry out the research, the methodology used consists of a case study with quantitative and qualitative evaluation, bibliographical research, through the theoretical foundation of production processes and costs, agribusiness management, evaluation of results. The results were obtained by comparing the results of each production process in relation to its investment, which was made possible by obtaining the data provided by the controls already in place on the property. The data was also used to calculate the production costs of each production activity, and through the cost of production and the value of the products sold it was possible to obtain the profit and profitability of each production process, and with the total value of the investment it was possible to analyze the profitability of the productions and compare their results. The property's two productive activities, soy and milk, represent a greater return for the property and the owner than the business alternative of renting. However, it was possible to analyze the results and decide on the best business alternative for the rural property, providing recommendations and conclusions on the subject, with a view to finding the best business alternatives and helping with future decision-making on the property.

Keywords: production process - costs - profitability

INTRODUCTION

Management tools are becoming part of and changing the decision-making structures on farms, where these changes end up affecting and being demanded, requiring management controls and the need to keep up to date with technological developments in order to increase productivity.

As productivity increases, business alternatives emerge and need to be analyzed further in order to be effectively implemented. It is necessary to analyze the agricultural production that will bring the most return to the property, or whether the business alternatives desired as minimum results can be obtained with the rural property's current production.

Within this context is the rural property that is the subject of this study, which aims to evaluate the most profitable agricultural products in order to identify the production process that will bring the greatest return to the property. In this sense, the study includes four chapters, developed with the aim of contributing to the foundation of the best business alternative.

Initially, the work is contextualized, with the delimitation of the theme, the justification and the problem, the general and specific objectives and the methodology applied in the research to achieve the proposed objectives, through data collection and analysis and interpretation, complemented by the presentation of the organization that is the object of this study.

Next, the theoretical framework is presented, which deals with production processes and their costs, costs of agricultural production processes, depreciation, costing methods, agribusiness management, internal and agricultural controls, evaluation of investment results, expected returns and indicators for evaluating results.

The next chapter deals with the diagnosis and analysis of the proposed objectives, describing the production processes of each activity on the property, as well as identifying the costs of the processes, the profitability of each production process and the analysis of business alternatives. The procedures were developed using data from the property's internal controls and the procedures using electronic spreadsheets.

The next stage presents recommendations and suggestions for the property, which are important for analyzing alternatives and making future decisions for the property.

This research concludes by presenting the results obtained at each stage of development and the proposed objectives. The results are presented in relation to the best business alternative for the rural property, which contributes to the development and growth of the property with products that add greater value.

Chapter 1

1 CONTEXT OF THE WORK

The aim of this chapter is to present the organization of the work, with a presentation of the rural property that is the subject of this study, the data collection and interpretation plan, the methodology defined for the research to achieve the defined objectives, which are the general and specific objectives, the justification and the problem question of this study.

1.1 THEME

Analysis of the profitability of agricultural production on a small rural property.

1.2 DELIMITATION OF THE THEME

Analysis of the profitability of soybean production in the 2011/2012 and 2012/2013 harvest periods, milk production from May 2012 to April 2013 and land leasing on a small rural property in the municipality of Santa Rosa, RS.

1.3 PROBLEM

Agriculture in Brazil has been undergoing positive transformations for the country's economy, including the evolution of the way people work in rural areas. The transformations in work are based on the introduction of technologies such as machinery and equipment that help to increase production, which supports the generation of food for the gradually growing population.

Among the technological means that can be found in all areas of agricultural production are the equipment and processes for preparing the land, for planting, for guaranteeing production and for harvesting, speeding up production processes. Most of these processes are carried out with the help of professionals, technicians in the field who are looking for the best way to increase production in the right way without mistreating the soil, using manure and fertilizers so that the land produces more with a greater return for the owner.

In this context, the farms are helped by investments, credits that the government provides at low interest rates to help family farming. They also need management techniques to help them make decisions.

In order to manage their property well, owners need to be aware of the costs that occur in production, which involve the prices paid for the inputs consumed in production, knowing what the market is looking for and the price of the final product paid by the market.

It is very important for rural producers to know the business area in which they operate well, and to know the market, climate and soil conditions, so that they can obtain better production results.

As such, it is clear that agriculture needs forms of management that help the owner make decisions. According to Conab (2010), the cost of agricultural production is an important control and management tool, important for decision-making. It is necessary to master the technologies and the results of spending during the productive phase of the crop.

For agriculture, the essential thing is to have control of your production costs, to check or alternatives for this investment. Owners need to be aware of profitability in order to make decisions about the direction of the business, if the expected results are not what they expected.

Among the agricultural productions mentioned above, the Northwest region stands out for its soybean cultivation and dairy farming, both of which have companies that buy their produce from rural landowners.

Thus, the following problem question arises: Is the minimum result desired by the entrepreneur obtainable with soybean and milk production?

1.4 OBJECTIVES

The objectives determine what you want to achieve during the research project and the results you want to achieve. According to Silva and Menezes (2001, p. 31): "The objectives will inform what you are proposing the research for, that is, what results you intend to achieve or what contribution your research will actually provide."

The objectives are divided into general and specific. "The general objectives define what the researcher intends to achieve with their research". (MEDEIROS, 2010, p.222), and the same author states that the specific objectives detail the general objective, defining the stages of the work (2010).

1.4.1 General Objective

Evaluate the most profitable agricultural products on small rural properties in order to analyze alternatives.

1.4.2 Specific objectives

    a) Describe production processes

    b) Identify the costs of production processes

    c) Determine the methodology for controlling the recording of expenses

    d) Evaluate the profitability of soy and milk production by comparing the results.

    e) Compare the alternatives for producing and renting the property.

1.5 BACKGROUND

Agriculture in Brazil has undergone many transformations. One of the transformations that has been taking place in a positive way is the use of technologies in the rural work environment, which supports the increase in productivity and also supports the reduction of the rural workforce.

With the increase in technology and productivity, it is extremely important for farmers and rural managers to have a broad knowledge of what is happening with their production, with regard to the quantitative and qualitative assessment of the economic and financial situation that this production brings to the property.

It is necessary to know the real situation of what has been happening on the property in order to make the right decision in choosing the crop to invest in so that it brings the owner the minimum return he expects.

In this way, the research project on the profitability of production on a rural property aims to guide the owners in their future decision on the best crop and/or production to be planted on the property and which will offer them a return.

The project is justified because it aims to assess the real situation of their production and sales, whether what they are producing is bringing them a return and whether it is viable for the property to continue producing. This study aims to help with future decisions on the property.

The question of study is extremely important for the academic, as it will improve his knowledge, his academic training and in the future will have an influence on his professional career, being able to apply this knowledge to other companies, and/or rural properties.

It is considered relevant to the Faculdades Integradas Machado de Assis Teaching Institution, as it is located in a region with an agricultural economy, and this study will contribute to the studies of other academics in this area, adding knowledge.

1.6 METHODOLOGY

The methodology depends on the objectives and research options used to carry out the work. According to Silva (2008, p.53): "The methodology to be employed in a research must be carried out from the formulation of the problem, the hypotheses raised to the delimitation of the universe or sample."

Methodology assists in the preparation of scientific work, helping in the way research is carried out, according to each area of work. Research methods provide the correct way to carry out research, helping scientists to prepare their work.

For Marconi, Lakatos (2010, p.65):

> [...] the method is the set of systematic and rational activities that, with greater certainty and economy, allows the objective to be achieved - valid and true knowledge - by tracing the path to be followed, detecting errors and aiding the scientist's decisions.

According to Cervo, Bervian, Da Silva (2007, p.27): "[...] the method is the order that must be imposed on the different processes necessary to achieve a certain end or a desired result".

According to Silva (2008), methodology studies the methods that are used to prepare work, to search for knowledge in certain areas.

Methodology in scientific work helps in the way the research is carried out and in the design of the entire study. It identifies the type of research that must be carried out in order to obtain the desired result.

According to Furasté (2008, p.33): "Another relevant aspect in the collection of material, is the establishment of the type of research that will be used to search for the information necessary to compose the knowledge that is desired."

In this way, it can be said that methodology means procedure, how to do and develop research.

1.6.1 Research Categorization

The internship report is based on applied research, which is aimed at acquiring knowledge with the aim of applying it to a specific situation (GIL, 2010).

Scientific research, during its development, has many methods inserted into its context. In order to carry out scientific research, it is necessary to know its techniques. According to Medeiros (2010) the research techniques are documentary and bibliographical research.

According to Medeiros (2010, p.35):

> Documentary research involves collecting documents that have not yet been used as a basis for research. Documents can be found in public archives, or those of private companies, in the archives of organizations and/or scientific institutions, in the archives of religious institutions, or even private ones, in registry offices, museums, video libraries, film libraries, correspondence, diaries, memoirs, autobiographies, or collections of photographs.

Documentary research is carried out on materials that have not yet been analyzed (SILVA, MENEZES, 2001).

According to Marconi, Lakatos (2010, p.166): "Bibliographical research, or research of secondary sources, covers all bibliography already made public in relation to the subject of study, [...]".

According to Gil (2010), most bibliographical research is carried out in printed publications, but with changes in technology, bibliographical research can be carried out in files published on the Internet, recorded on CDs, among others.

Bibliographical research will help the author to prepare his study with coherent statements in accordance with his object of study. According to Medeiros (2010, p.39) the aim of bibliographical research: "[...] is to place the author of the new research before the information on the subject of interest".

In this context, this study used documentary and bibliographical research, using books, newsletters and property documents to obtain the necessary data.

As for the method, this study is quantitative, as research was carried out into the production cost controls of the property analyzed. According to Silva (2008, p.28) "it means quantifying opinions, data, in the form of data collection."

Quantitative research requires statistical data to be collected and processed. Quantitative research provides

relationship between variables and causality between phenomena. (BEUREN et al, 2004)

The same authors go on to complement qualitative research as:

> Qualitative research involves more in-depth analysis of the phenomenon being studied. The qualitative approach aims to highlight characteristics not observed through a quantitative study, given the superficiality of the latter. (BEUREN et al, 2004, p. 92)

In this way, the study uses quantitative and qualitative methods, and a case study was used to support the elaboration of the research project, which will delve deeper into some issues through reports available from the analyzed property. The case study "consists of the in-depth and exhaustive study of one or a few objects, in such a way as to allow its broad and detailed knowledge [...]" (GIL, 2010, p.37).

According to Furasté (2008, p.37): "In this type of research, an exhaustive study is made of some particular case, person or institution, in order to analyze the specific circumstances surrounding it."

Based on these techniques, methods and types of research, management analysis is carried out, involving the profitability of the crops produced, based on their current production processes and possible alternatives.

1.6.2 Data collected

The data collection procedure begins with the implementation of the types of instruments and techniques selected to carry out the data collection. (MARCONI, LAKATOS, 2010)

Data was first collected through bibliographies, documents available for research, which will serve as the basis for the research project. Documentary consultations are necessary for a good case study (GIL, 2010).

According to Marconi, Lakatos: "There are various procedures for collecting data, which vary according to the circumstances or the type of researched." (2010, p. 149).

Secondly, for this study, data was collected from documents, cost spreadsheets and controls of soybean production from the 2011/2012 and 2012/2013 harvests, and milk production from May 2012 to April 2013 on the property under study.

1.6.3 Data analysis and interpretation

After the data was collected, it was tabulated in spreadsheet form in order to allow for its interpretation.

According to Marconi and Lakatos (2010, p.152), interpretation "is the intellectual activity that seeks to give greater meaning to answers by linking them to other knowledge."

The analysis of the data collected involves the results obtained and the appropriate comparison with the results proposed or previously published in theories or practiced studies. For Gil:

> The process of analyzing the data involves various procedures: coding the answers, tabulating the data and statistical calculations. After, or together with, the analysis, there can also be data interpretation, which basically consists of establishing the link between the results obtained and others already known, whether derived from theories or from studies carried out previously (2010, p.113).

Data analysis tries to relate the phenomenon studied to the analysis of different contents (MARCONI, LAKATOS, 2010).

According to Silva and Menezes (2001, p.35): "Analysis should be carried out to meet the research objectives and to compare and contrast data and evidence with the aim of confirming or rejecting the hypothesis(es) or research assumptions."

By analyzing and interpreting the data, conclusions can be reached in order to reject or accept the analyses, according to the research assumptions. The main conclusions were drawn from the results of the 2011/2012 and 2012/2013 soybean harvests and the monthly results of milk production from May 2012 to April 2013.

This research was based on collective data from spreadsheets that allow for the analysis of results, profits and returns. Finally, an internship report was drawn up concluding the study and comparing the results.

1.6.4 Presented by Organized

The rural property, which is the subject of this study, is located in Linha Sete de Setembro Norte, Santa Rosa/RS. It has an area of 12.5 hectares of land, divided into parts, with 2 hectares set aside for the preservation of native forest; 2 hectares of farmland, 0.5 hectares of land for housing, patio and orchard, and the remaining 8 hectares set aside for farming, planting soybeans, pasture and corn for dairy production.

The rural property was acquired in May 1994 from the Mayer family by the current owner, Mr. Edgar Eberhardt, who came with his family to work on their own crops and earn a living for his family.

The owner is responsible for management, with the help of his wife and daughters.

The farm's source of income is milk and soya. The milk is produced continuously and sold every two days, with the proceeds received monthly. The main customer is Laticínios Tchê, located in Santo Cristo/RS, and the company that collects the milk is Transportadora Igo Ltda.

As for soybeans, the customers are Dinon Cereais, which has a branch in Linha Sete de

Setembro and its head office in the municipality of Tuparendi/RS, and Cooperativa Cotrirosa, which has its branch in Vila Sete de Setembro. The suppliers of inputs for soy and milk production are Cooperativa Mista Sao Luis Ltda. and Comercial Dinon Cereais.

The property is located in front of the Sete de Setembro North Line road, with Felipe Carli Steffan's property as its neighbors to the left, at the end of the property it borders Arlindo Streich's land, and to the right is Gelson Hein's property, which can be seen in the following image.

Below is a map of the property using an image available on Google Earth.

Illustration 1: Photo of rural property
Source: Adapted from Google Earth (2005)

The area of the property that is the subject of this research is differentiated, with a black outline, from the other properties. The image was taken in 2005 (two thousand and five). This image can be seen even a few years after the update, as the property still has the same geographical features and production style.

Chapter 2

2 THEORETICAL FRAMEWORK

The theoretical framework deals with theories that complement the objectives and data collected in order to better understand the subject and analyze its conclusions. Its aim is to improve knowledge, provide a basis and help with the course completion work.

The theoretical framework is made up of research on the subjects of production processes, costs of production processes, agricultural and farming costs, agribusiness management, internal control, control of production processes in agriculture, evaluation of results, evaluation of results in agribusiness, investments, expected returns and indicators for evaluating profitability in agribusiness, all of which were covered in order to facilitate the study and the objectives of this work.

2.1 PRODUCTION PROCESSES

As a concept of production processes, Vieira says: "The production sector, also called industry, factory, production cells, is the place in the company where the processes of transforming materials into final products are carried out." (2008, p. 54).

> A production process is any integrated group of activities designed to achieve a specific purpose. All processes use physical elements, such as people and equipment. Some processes have tangible outcomes, such as completing or delivering a unit of production; some processes have intangible outcomes such as customer or employee satisfaction. (ATKINSON et al, 2011, p. 579)

The flowchart technique is used to describe the processes (CRUZ, 2011). Production processes transform materials into end products, and it's the same in rural areas. In rural areas, production processes involve soil preparation, planting and harvesting, all depending on the type of crop and production chosen on the rural property. Crepaldi (2009) states that production processes in the agricultural area involve soil preparation, preparation for planting, planting, harvesting and harvested products.

Rural production systems are widespread throughout the country, with processes and production changing according to each region and depending on each producer (CANZIANI, 2009).

According to CANZIANI (2009, p.17): "There are many existing production systems in the country, with significant variations between regions and between producers[...]". Production systems are developed according to the characteristics of each region.

In this context, it can be seen that the production process consumes resources in organizations, which are considered to be costs, expenses and even investments, necessary for its realization.

2.2 PRODUCTION PROCESS COSTS

Production processes meet consumer needs by transforming materials and resources into goods and services. These production processes have factors that add to their transformation, including production costs. According to Siedenberg and Pasqualini (2009, p. 12):

> Production Management is the activity of managing scarce resources and processes that produce and deliver goods and services in order to meet the needs and/or desires of its customers in terms of quality, time and cost.

In order to understand production costs, it is necessary to know the terms used, which are: Expenses   , Costs, Expenditures, Investments (MEGLIORINI, 2012).

The expense is realized when the acquired asset becomes the property of the company; costs are all the expenses used in the production process; Expenses are expenses not used in the production process, which go directly to the income statement; Investments are expenses intended for the benefit of future periods in the company. (VICECONTI, NEVES 2013).

Production has process costs, which are divided into direct and indirect, depending on each production.

According to Pereira (2009, p. 04): "Indirect production cost center: These are the cost centers that must be apportioned (distributed) to productive cost centers." And direct production costs: "(...) are the cost centers associated directly with manufacturing and are usually defined according to certain criteria." (PEREIRA, 2009, p. 04).

In relation to production levels, there are fixed and variable costs. Fixed costs are independent of production, while variable costs occur according to production. "Fixed costs are those that result from maintaining the company's production structure, regardless of the quantity that will be manufactured within the limits of the installed capacity." (Megliorini, 2012, p.11). The same author states that: "Variable costs: are those that increase or decrease according to the volume of production." (2012, p.11).

For Rizzon:

> It is all too common to consider that the production costs of a given good are made up exclusively of the raw materials, packaging and other direct materials used in its manufacture, forgetting to add other costs to these prices, such as direct labor and fixed costs. (2006, p. 10)

The cost of production is defined according to the production system of each process. "The main purpose of the process or continuous costing system is to determine the costs and expenses of the production cycle" (VELTER, MISSAGIA, 2010, p. 146). Process costing is used when there is continuity in the production of the same or similar products (IUDÍCIBUS, 2009).

Production processes that have continuous costs are accumulated during the processes and divided by the quantity produced after the end of the process, each month, week or production period. According to Martins (2008, p.145) continuous production has:

> Their costs are accumulated in accounts representing the various production lines; these accounts are always closed at the end of each period (month, week, quarter or year, depending on the company's minimum cost accounting period).

Martins (2008, p 145) also states that: "[...] in the calculation by Process, costs are not evaluated unit by unit, but on the basis of the average cost for the period (dividing the total cost by the quantity produced)."

> [...] the concern of Cost Accounting is to determine and control costs by departments, by sectors, by production phases (processes) and then divide these costs by the quantity of products manufactured in the process, during a certain period - to cost the manufacturing process in a given period (VIEIRA, 2008, p.54)

For each period, you have all the production costs and you need to divide them by the number of products manufactured to obtain the unit cost of each unit produced.

Make-to-order costs arise at the start of an order, a production order, and end when the production process is completed. "Costs in production by order are recorded in a specific account for each order, which only stops receiving costs when the service or production is completed." (VELTER, MISSAGIA, 2010, p.144).

Production order costing has the total costs divided into: "[...] direct material, direct labor and support costs, estimated or identified with an order [...]" (ATKINSON et al, 2011, p.250).

In order to reduce production costs, it is necessary to evaluate production capacity and reduce some of the factors that make up production. Conab (2010, p. 14) states that:

> When deciding what and how much, how and for whom to produce, taking into account consumer responses, companies try to vary the quantity of factors used, in order to vary the quantity of product produced. In this process, they always seek to use the best technology at the lowest cost.

Production processes include the use of technologies that help improve production, the use of inputs that can change the costs of a given production.

> The production function represents the technology used in the production process of a given product and the technology determines which inputs, their quantity and how they are used. Given a production technology, the prices and quantities of inputs will determine the total costs and in view of the different possibilities for using these factors, it is possible to combine them in such a way as to minimize production costs. (CASTRO et al, apud

CONAB, 2010, p.15)

According to Iudícibus (2009), the cost of the product refers to the inputs used in finished production, and the costs of the period only occur when the products are sold.

By knowing all the costs of the production processes, the profitability of the business is obtained and the manager is then able to analyze whether his business is bringing in the desired return. To do this, you need to know the opportunity cost of your business. According to Assaf Neto (2010, p.82), opportunity cost "refers to the return on the best financial alternative available on the market which an investor has given up in order to invest in another".

In this context, there are also the costs of agricultural and farming production processes and it is worth highlighting their importance for rural production.

2.3 AGRICULTURAL AND FARMING PRODUCTION COSTS

Rural costs have long been treated as unnecessary for the business. The issue that has been changing the real situation is the importance of knowing the costs to help make decisions on the property and "[...] one of the most fundamental pieces of information for the proper management of the Rural Company: the operating costs." (CREPALDI 2009, p.151)

The costs of agricultural production processes are extremely important because they give rural managers the ability to analyze and manage their production more effectively. CONAB states that:

> The cost of agricultural production is an exceptional tool for controlling and managing production activities and generating important information to support decision-making by rural producers (2010, p. 08).

Costs provide important information to help make decisions. Cost information is used to decide on the type of investment to be made in the plantation. According to Conab: "The results of production costs are directly related to the cultivation systems and agricultural model adopted by the rural producer." (2010, p.11)

Agricultural production costs allow producers to know their production costs, profitability and the break-even point of the crop produced. According to Crepaldi (2009, p.152), controlling costs has advantages: "It also allows the rural entrepreneur to know the profitability of his business and determine his company's break-even point." The same author states that with costs, the owner can analyze the causes of success or loss in the company, with the ability to change this scenario by increasing profits or correcting the errors that lead to losses (2009). "It is essential that production costs are also seen as an instrument for improving the management of the modal production unit, and can be one of the variables in increasing the rural producer's income." (CONAB, 2010, p.27).

The costs of culture are different from the expenses of the period. Marion says that crop costs are:

> [...] all expenses that can be identified directly or indirectly with the crop (or product), such as seeds, fertilizers, labor (directly or indirectly), fuel, depreciation of machinery and equipment used in the crop, agronomic and topographical services, etc. (2010, p.15)

For Marion: "Expenses for the period are understood to be all expenses that are not identifiable with the crop and are therefore not accumulated in stock (temporary crops), but are appropriated as expenses for the period." (2010, p. 15)

Costs in family farming need to be reduced in the same way that large farms reduce costs by implementing new technologies. According to Andrioli (2008, p.02): "For a family farmer to be able to reduce the working time needed to produce something, it would mean that he would have to unemploy himself, his children or someone in his family."

The same author emphasizes the importance of smallholdings and family farming having several activities that bring results, not depending on just one activity on the property to support the family. (ANDRIOLI, 2008)

Farming costs require a complete system in order to manage costs/revenues correctly, and the characteristics of each farm must be taken into account.

> It is assumed that some of these uncertainties and threats can be resolved with the development of a management system for agricultural planning and costs,

through which the producer or rural entrepreneur can manage their livestock activities (breeding, rearing, fattening, milk, etc.), agricultural activities (grains, fruit growing, horticulture, etc.), and any other rural activity, in an integrated or independent manner. (MARION, SEGATTI, 2006, p.04)

The components of costs from the point of view of measuring opportunity costs are divided into: explicit costs and implicit costs. Explicit costs occur when the values can be measured, they are expenses disbursed by the producer in the course of production activity; implicit costs are not directly disbursed in the production process, they correspond to the remuneration of fixed capital and land and depreciation expenses. (CONAB, 2010)

According to Araújo, Araújo, Correia (2008, p.02), agricultural costs are grouped into two categories:

Effective operating costs (COE), which correspond to the variable costs and direct expenses with financial outlay from soil preparation to harvest, and indirect costs (CI), which reflect the fixed costs and indirect expenses incurred by the producer to obtain production, such as the cost of the land, depreciation, the foreman's salary, taxes, etc.

Production costs are not always the same in farming. The variables are the type of production and the crop to be produced. According to Nogueira (2009), the costs of the products produced are divided into costs per production order and costs per process.

The structure of total production costs in farming provides information about the product, its productivity and the characterization of fixed and variable costs. Variable costs are divided into: machinery and expenses, implements and utensils, maintenance of improvements, temporary labor, inputs, overheads, external transport, storage and financial charges. And the fixed costs are: depreciation; remuneration on non-depreciated equity - rate of return; insurance, fees and taxes; fixed labor (for the administrator); land remuneration (CONAB, 2010).

Agricultural production costs are grouped and separated into the cost of farming, post-harvest expenses and financial expenses, which are the variable costs. Fixed costs are represented by depreciation, other fixed costs and factor income. According to Conab, the description of agricultural costs is shown in illustration 2.

Description of the items that make up the cost of production:

A. VARIABLE COSTING

1- FARMING COSTS

1- Operating machinery and implements

2- Labor and social and labor charges

3- Seeds

4- Fertilizers

5- Pesticides

6- Irrigation costs

7- Administrative expenses

8- Other items

II- POST-HARVEST EXPENSES

1- Agricultural Insurance

2- External transportation

3- Technical assistance and rural extension

4- Storage

5- Administrative Expenses

6- Other items

III- FINANCIAL EXPENSES

1- Interest

B - FIXED COST

IV- BELITTLEMENT AND EXHAUSTION

1- Depreciation of improvements and installations

2- Machinery depreciation
3- Depreciation of implements
4- Crop exhaustion
V- OTHER FIXED COSTS
1- Labor and social and labor charges
2- Fixed capital insurance
C- OPERATING COST (A+B)
VI- FACTOR INCOME
1- Expected return on fixed capital
2- Terra
D- TOTAL COST (C+V)

Illustration 2: Description of the items that make up agricultural costs.

Source: Adapted from Conab (2010, p.29)

Agricultural costs can be considered for soybean, corn, wheat and other crop activities that have common costs.

The costs of dairy farming are broken down into variable and fixed costs.

SPECIFICATION	R$ LITER	%
1. VARIABLE COSTS OF DAIRY FARMING		
Manpower for herd management	0,05996	18,882
Concentrates	0,05157	16,240
Minerals	0,00440	1,386
Green fodder	0,0391	9,732
Silage	0, 00729	2,295
Medicines	0,01392	4,384
Artificial Insemination	0,00917	2,888
Milk transportation	0,02450	7,714
Energy and fuel	0,00648	2,039

Social security contribution	0,00710	2,235
Repairs to improvements	0,00510	1,605
Repairs to machinery, engines and equipment	0,00440	1,387
Remuneration of working capital	0,00580	1,825
TOTAL VARIABLE COSTS OF DAIRY FARMING	**0,23060**	**72,611**
2. FIXED COSTS OF DAIRY FARMING		
Depreciation	0,04403	13,864
Taxes and fees	0,00444	1,398
Return on Fixed Capital	0,03851	12,127
TOTAL FIXED COSTS OF DAIRY FARMING	0,08698	27,389
3. TOTAL COST OF DAIRY FARMING (1+2)	0,31758	100,00

4. SELLING ANIMALS	0,07259	
5. TOTAL COST OF MILK (3-4)	0,24498	
VARIABLE MILK COSTS	0,17789	
FIXED MILK COSTS	0,06710	
TOTAL COST OF MILK	0,24498	

Illustration 3: National Milk Production Costs in the Central, South and Southeast Regions
Source: Adapted from Crepaldi (2009, p.167)

The costs listed in illustration 3 represent the costs of dairy farming that can be applied to the southern region.

It can be seen that the biggest variable cost is labor, which accounts for eighteen percent of total costs, followed by concentrates and green fodder. The biggest fixed costs are depreciation and remuneration of fixed capital.

With the exposure of costs, we realize the need for control and management in agribusiness.

1.1.1 Costing methods

Costing methods help to calculate product costs, separating costs according to the decision and best benefit for each company. "The traditional forms of costing are: absorption costing, direct or variable costing, standard costing, RKW, and ABC." (VELTER, MISSAGIA, 2010, p.122).

Absorption costing is aligned with the principle of competence and attributes all direct and indirect costs to the product and/or service. It is used for management and tax purposes (VELTER, MISSAGIA, 2010).

Variable costing is used for management purposes and the costs used in production are allocated to products. According to Vieira (2008, p.63). "The Variable or Direct Costing method is used for management purposes, providing tools that help in the company's management process."

Variable costing allocates only the variable costs, leaving the fixed costs only for the Results, "[...] considers only the direct or variable costs as manufacturing costs". (RIBEIRO, 2009, p.57), and the same author emphasizes that direct costs vary according to the volume of production. (2009)

> Variable costing [...] is a type of costing that consists of considering only the variable costs incurred as the cost of production for the period. Fixed costs, because they exist even if there is no production, are not considered to be production costs but expenses, and are charged directly to the income statement for the period. (VICECONTI, NEVES, 2013, p. 131).

Variable costing is used as a management tool, providing the contribution margin of each product to the company's results, according to Velter, Missagia (2010, p. 128): "[...] this method does not provide the cost value of a product, but determines the contribution that each product brings to the company in the form of results."

By using variable costing you can see the contribution that each product makes to the company, a contribution that affects the result, either positively or negatively. With the contribution margin, you can see which product brings the most positive results and advantages. According to Velter and Missagia (2010), the contribution margin is calculated as the difference between revenue and variable costing for each product.

The RKW method applies apportionment of all costs and expenses to its products. Standard costing has a parameter for costs and must adjust them to the actual cost incurred. And ABC costing uses several indirect cost centers arising from various activities, seeking to distort apportionments caused by the other costing methods (VELTER, MISSAGIA, 2010).

In ABC costing, "a company's resources are consumed by its activities and the products emerge as a consequence of the activities considered strictly necessary to manufacture or market them." (VALLE, OLIVEIRA, 2012, p.133-134). In this costing method, indirect costs are assigned to products according to the activities and consumption carried out by each production.

1.1.2 Depreciation

Depreciation refers to a fixed amount that occurs due to the wear and tear of the asset used, regardless of the process. According to Szuster et al (2008, p.326): "[...] depreciation corresponds to a reduction in the value of fixed assets."

The reduction in the value of fixed assets occurs when there is wear and tear due to use, the action of nature and technological obsolescence. (SZUSTER et al, 2008), and "the entity must allocate the depreciable value of assets on a systematic basis over their useful life." (NBC T 19.41, 2011, p.121). Depreciation is part of the cost of production and refers to the loss of value or productive efficiency, representing a real cost (CONAB, 2010).

For Hendriksen, Van Breda (2012, p.324) it is necessary to "[...] apply a standard or a formula for allocating the original or corrected value (less residual value) to period expenses or product costs." (ATKINSON et al., 2011). To calculate annual depreciation, the historical cost is used with the difference of the residual value and the result is divided by the useful life of the asset. (ATKINSON et al, 2011).

With depreciation, the asset reduces its value over time of use, over the years of its useful life, reducing its productive capacity. The depreciation of breeding and producing animals "must take into account the residual value of the animal, which can be obtained by selling it after it has been used for breeding and/or work." (CREPALDI, 2009, p.134).

Depreciation occurs as assets are used in production processes, and the same is true of production units in agribusiness.

2.4 AGRIBUSINESS MANAGEMENT

Management in companies is extremely important in order to achieve goals and results for the organization. Agricultural activity requires investment in land and machinery in order to continue producing.

> In agriculture, managing a rural enterprise requires technology and knowledge to deal with risks and uncertainties specific to the sector (climate, politics, economy, legislation, etc.), the instability of income due to productivity and internal and external prices, the characteristics of oligopoly and oligopsony in commerce and industry that relate to agriculture, variations in prices and marketing difficulties during the harvest, often problematic credit, the perishability of agricultural products, as well as the very complexity of agricultural production (location, time, space, climate, environment, soil, etc). (CONAB, 2010, p. 13)

The organization needs to be aware of the factors of production that it will use in its production. There needs to be a concern about grouping together all the resources and factors of production available for the rural property to function efficiently (CANZIANI, 2009).

In this way, the owner also needs to manage the information relating to the business, in order to be aware of the total invested and produced. With this information, the owner is aware of the expenses, outflows and inflows that occur for production. Crepaldi states that:

> Management information is the result of what actually happens on the farm.

By classifying and organizing the data relating to the daily economic and financial movement of the property, it is possible to generate this information. They will indicate the volume of income per activity, the levels of investment per sector and the amounts disbursed. (2009, p 55)

The use of management information is fundamental to the running of a business. What often happens is the lack of information and professional training on the part of the owners, who face difficulties in their management.

According to Neukirchen, Zanchet, Paula (2005, p. 07):

For agriculture, the reality is even more painful, since access to information that allows producers to become professional can be considered the main obstacle faced by the class, in addition to cultural factors that hinder its development.

According to Crepaldi (2009, p.2): "(...) in the current situation of the farmer's linkage and dependence on the market, in-depth knowledge of the business becomes indispensable for rural producers."

With knowledge of the business and the activity to be produced, it is up to the rural administrator to decide how to use natural resources. According to Crepaldi (2009, p.2): "It is up to him to decide what, how much and how to produce, to control the action after starting the activity and, finally, to evaluate the results achieved and compare them with what was initially planned."

Rural property is considered an organization and needs to be managed, just like a company.

The organization of the agricultural enterprise is understood as the combination of activities carried out according to the characteristics of the available factors of production. This means choosing all the crops and livestock that will be farmed in order to make the best possible use of the land, improvements, machinery and implements and labor. (CREPALDI, 2009, p.6)

In order to achieve good farm management, it is necessary to obtain the correct data to generate coherent information for the manager,

[...] by providing information that enables planning, control and decision-making, transforming farms into companies with the capacity to keep up with developments in the sector, especially with regard to the objectives and duties of financial management, cost control, crop diversification and comparing results. (SEGALA, SILVA, 2007, p. 62)

Management in agribusiness requires rural managers to implement production administration and also to control all their production, i.e. property management. According to Canziani (apud Canziani 2009, p.207): "One way of evaluating the management or administration of an agricultural company can be done with the use of a matrix of administrative activities."
The same author comments that:

This matrix links the functions of planning, organizing, directing and controlling, carried out at the strategic and operational level of the agricultural company, with the administrative areas of production, finance, marketing and personnel, thus making up the administrative process. (2009, p.207)

Management in agribusiness is complete with the functions carried out at the strategic and operational levels, organizing the company so that business runs smoothly.

In this context, it is important to detail the procedure for effective control of the property and its production.

2.5 INTERNAL CONTROLS

All companies need internal controls to keep track of their activities. With good internal control, companies can achieve the best results, with less waste. (Crepaldi, 2012) These controls exercise control over production. "Among the various types of control that can be carried out in the production of goods and/or the provision of services, we can mention the control of deadlines, material, costs, quality and work." (PASQUALINI, LOPES, SIEDENBERG, 2010, p.83)

Each company has controls according to each activity carried out, to keep procedures better organized according to each activity.

> In an organization, internal control represents the set of procedures, methods or routines with the aim of protecting assets, producing reliable accounting data and helping management to conduct the company's business in an orderly manner (ALMEIDA, 2008, p.63).

Control is essential for the smooth running of the company's business and for production control. "Control is the set of methods and tools that company members use to keep the company on track to achieve its objectives." (ATKINSON et al, 2011, p. 581)

The control of companies is embedded in the processes that are going to be carried out, in order to obtain better results with the control of each production. Internal control is therefore related to the need for information. (CREPALDI, 2012). "It is of fundamental importance to use adequate control over each operating system, as this is how the most favorable results are achieved with the least waste." (CREPALDI, 2012, p. 69).

By implementing control in their operational activities, companies obtain information on production costs, their results and investments, and can decide on the process that brings the most advantage to the organization.

Internal controls can be applied to different sectors and responsibility centers, which are: cost centers, profit centers and investment centers (HORNGREN, SUNDEN, STRATTON, 2004).

Internal control is applied in different sectors and centers of responsibility, and it is necessary to demonstrate the need for controls in rural activity.

2.5.1 Control of Production Processes in Agribusiness

In farming, control is important in providing the producer with the right amount of inputs consumed, the total context of their production costs, and the amount of stock for sale. According to Ribeiro, you need to have

> [...] the cost-benefit ratio of maintaining more complex controls in search of more accurate information must always be considered. On the other hand, the need for controls, however simple they may appear to be, becomes an indispensable factor when it comes to valuing agricultural stocks. (2004, p. 211)

Marion apud Marion, Segatti (2006) states that there needs to be integrated cost control, control of machinery, depreciation calculations, total operating costs and unit costs to help make decisions.

Therefore, internal control of inputs, labor and indirect costs incurred in each production is important for rural properties, providing managers with correct information for decision-making.

In some cases, systems are used to carry out controls.

> A good internal control system is essential in an agricultural company, enabling the manager to monitor the development of each stage of the crop, as well as checking the allocation of the various inputs that make up these crops. (RIBEIRO, 2004, p. 216)

It is of the utmost importance for rural producers to have ample control of their businesses due to the growing competition in agribusiness.

> The growing need for greater efficiency in production processes within the agribusiness sector, mainly due to increased competition, makes it clear that it is important to have a control system that can provide essential information for understanding and improving the activities carried out by companies. (ALMEIDA, MACEDO, 2010, p.05)

The market imposes the price of your products, which is one of the relevant factors in implementing controls over your costs and the inputs used in your production.

Once effective controls are in place, it is possible to determine methods and procedures for evaluating and monitoring results.

2.6 . EVALUATION OF RESULTS

Companies and production units obtain their results through their sales minus the costs and expenses incurred during the production period, obtaining the net profit of the operation. Padoveze (2010) states that the results and expenses are obtained from the sum of the transactions that took place, which lead to the movements of the period analyzed.

The evaluation of results is related to the performance of each product. Iudícibus (2009) states that performance should not only take into account results, but also assess whether the company's sectors have achieved their goals.

> The main bases of comparison adopted for studying business results are total assets, shareholders' equity and sales revenues. The results normally used, in turn, are operating profit (profit generated by assets) and net profit (after income tax). (ASSAF NETO, 2010, p.109)

The results are evaluated according to the performance of each production, and with analysis of the investment made. In order to assess whether production has brought a return, it is necessary to have knowledge of the total investment, costs and profits. Results are evaluated according to return on investment, return on equity, degree of financial leverage, operating margin and productivity (ASSAF NETO, 2009).

Profitability is assessed using the data provided by the Income Statement, on Assets and Equity. The profitability ratios are: gross profit margin; operating profit margin; net profit margin; earnings per share; return on total assets; return on equity (GITMAN, 2010).

According to Velter, Missagia (2010) profitability indicators are represented by Operating Margin, Gross Margin, Operating Turnover, Asset Turnover, Operating Return, Return on Assets, Return on Equity, Degree of Financial Leverage, Degree of Operating Leverage.

Profitability ratios make it possible to visualize the situation of companies, which use the Income Statement, which helps to evaluate results, and also evaluates results in agribusiness.

2.6.1 Evaluating Results in Agribusiness

Agricultural activity needs production that brings results for the farm and its managers.

According to Creplaldi (2009), the difference between sales and the costs, expenses and investments paid for production is considered the result of rural activity.

In order to evaluate the results of rural activities, it is necessary to monitor the data and information obtained during the production of the activity. According to Crepaldi (2009, p.41): "The results management control system must be designed in such a way as to enable operational problems to be identified and the performance of each strategic business unit to be assessed."

> [...] as farms have small areas in which to develop their agricultural production, it is of the utmost importance that they develop them efficiently, in order to achieve not only sustainability, but also profitability in the activity they explore. (NEUKIRCHEN, ZANCHET, PAULA, 2005, p. 07)

Owners need to know the profitability of their production through collected data, production controls with costs and stocks produced.

> What managers need to know is how profitable their productive activity is. What are the results obtained and how can they be optimized by evaluating results, sources of income and types of expenditure? How can you improve income and reduce expenses? This analysis will only be possible once you know where you are spending resources and where you are generating income. (CREPALDI, 2009, p.55)

To make it possible to evaluate results, one of the fundamental items or points is to identify the amount of resources allocated to the activity. Among these, the following stand out as a starting point: investments

2.6.2 Investments

Investments represent the monetary value of the resources used to acquire goods, at the time the company acquires assets (ATKINSON et al, 2011).

The investments made represent the start of the activity, and profits are subsequently required in relation to the amount invested. Iudícibus (2009) states that profit must be related to the amount invested in order to achieve profitability.

According to Motta, Calóba (2009, p.34) adds that:

> An investment is a situation in which capital is invested in some way, whether in a new project, the purchase of an existing company, etc., with the aim of creating value, i.e. recovering the amount invested (principal), plus a return on the investment (interest rate), over a given period of time.

In rural activity, investments are made in production through the purchase of inputs, or even agricultural equipment, according to the planting period of each crop, or constantly when production is continuous.

According to Crepaldi (2009, p.332): "Investment is considered to be the application of financial resources, during the calendar year, aimed at the development of rural activity, the expansion of production and the improvement of the activity."

Investments are made to develop periodic activities and to expand and improve continuous production. Investments are made in: improvements, permanent crops, the purchase of machinery, working animals, technical services, inputs for production, roads for access to the property and production, installation (electricity), scholarships for training professionals (CREPALDI, 2009).

The investments made in Current Assets are called working capital because they represent "the continuous movement of the main elements that form the company's transactions (the company's business) where it basically makes its profits." (PADOVEZE, 2010, p.117-118)

With the investments made and/or planned, the owner estimates the return they want to expect from a given investment.

2.6.3 Expected Returns

The expected return is obtained from the cash flow generated by the investment in a given asset. According to Atkinson et al (2011, p.523) the return: "Is the growing cash flow in the future attributed to the long-term asset."

The return calculation is represented by profit before income tax and financial expenses divided by Total Assets (VELTER, MISSAGIA, 2010)." Analyzing the return on investment calculation is an excellent way of assessing the effectiveness of an enterprise." (VELTER, MISSAGIA, 2010, p.153):

The return on investment is generated by dividing the profit by the total investment. "Analysts calculate the accounting rate of return by dividing the average accounting profit by the level of investment." (ATKINSON et al, 2011, p.536)

> In general, therefore, we should relate the profit of an enterprise to some value that expresses its relative "size", in order to analyze how well the company did in a given period. The best concept of "size" could be sales volume, the value of total assets, operating assets, net worth, share capital, etc. All have their advantages and disadvantages (IUDICIBUS, 2009, p. 88).

The return by which the landowner can compare his production is the return based on renting the land. According to Crepaldi:

> A rural lease is a contract whereby one person undertakes to another, for a fixed period of time or not, the use and enjoyment of a rural property, part or parts of it, including or not other goods, improvements and facilities, with the aim of carrying out agricultural, livestock, agro-industrial, extractive or mixed activities on it, for a certain consideration or rent, subject to the percentage limits set out in the law.

With the rental value, the owner has the alternative of requiring the same return for his production.

The remuneration expected by the rural owner or producer should be at least 6% per year as a rate of return, in the same way as if the investment were made in an alternative business. (CONAB, 2010)

In the same way, the expected return for the land should be on its sale value. "[...] the land remuneration rate is 3% of the real average sale price of the land." (CONAB, 2010, p.44), and since land remuneration is a fixed cost, it should be calculated from 3 to 5% (three to five percent) of the value of the land or the value of the lease, whichever is the greater. (CONAB, 2010).

It is up to each rural manager to decide on the return that best suits their business, their production, deciding on the production and/or activity that brings the greatest return on their investment.

2.6.4 Indicators for Evaluating Profitability in Agribusiness

Profitability assessment indicators require a variety of information in order to provide the correct result. According to Crepaldi (2009), the property must have information on the balance sheet, cash control and stock. And evaluations must be compared with previous periods.

This data is used to calculate the indices. According to Crepaldi (2009, p.305): "(...) indices that reveal the degree of efficiency of management and the health of the business". The indices are: profitability index, profitability index and liquidity index (CREPALDI 2009).

According to Yamaguchi, Gomes, Carneiro (2006, p.01): "Profitability is a very important indicator for assessing the situation your business is in". Profitability represents the return on capital invested and the real situation of the business.

Profitability is assessed by the performance of each product produced on the farm. "The definition of performance [...] to be used is part of a logical sequence of procedures for developing and implementing a performance measurement and evaluation system." (ALMEIDA, MACEDO, 2010, p.07)

In rural activities, evaluating the performance and profitability of production is important, as it shows the reality of the property, evaluating the production and the alternatives that really add value to the property.

Chapter 3

3 DIAGNOSIS AND ANALYSIS

In order to analyze the profitability of agricultural production of soybeans from the 2011/2012 and 2012/2013 harvests, and milk from May 2012 to April 2013, with the alternative of leasing, this research develops the stages of their production processes of milk and soybeans, the identification of the resources necessary for the production of each activity and the expenses involved in each resource used and their due production costs.

With knowledge of the expenses required for each crop, you can analyze the profitability of the farm's crops in relation to the investment made in each crop. With this analysis, you will be able to see the best business and income alternative for the property to continue investing in. If agricultural production does not bring the desired profitability, there is the alternative of leasing, which is remuneration for borrowing the land, or for its production capacity.

The rural property analyzed is located in Linha Sete de Setembro Norte, in the interior of the municipality of Santa Rosa - RS, and its main income comes from dairy production and soybean cultivation. The owners had cost spreadsheets and production controls.

3.1 THE ORGANIZATION'S PRODUCTION PROCESSES

Two important production processes take place on the property, which are responsible for shaping the results of the rural activity. All the stages of each process are presented below.

3.1.1 Milk Production Process

The milk production process starts every day and takes place at the same time twice a day, in the early morning and late afternoon. At the beginning of the process, the cows are kept in a shelter, and then they are moved to the physical space where the milking takes place.

The following flowchart shows all the steps taken in the activity.

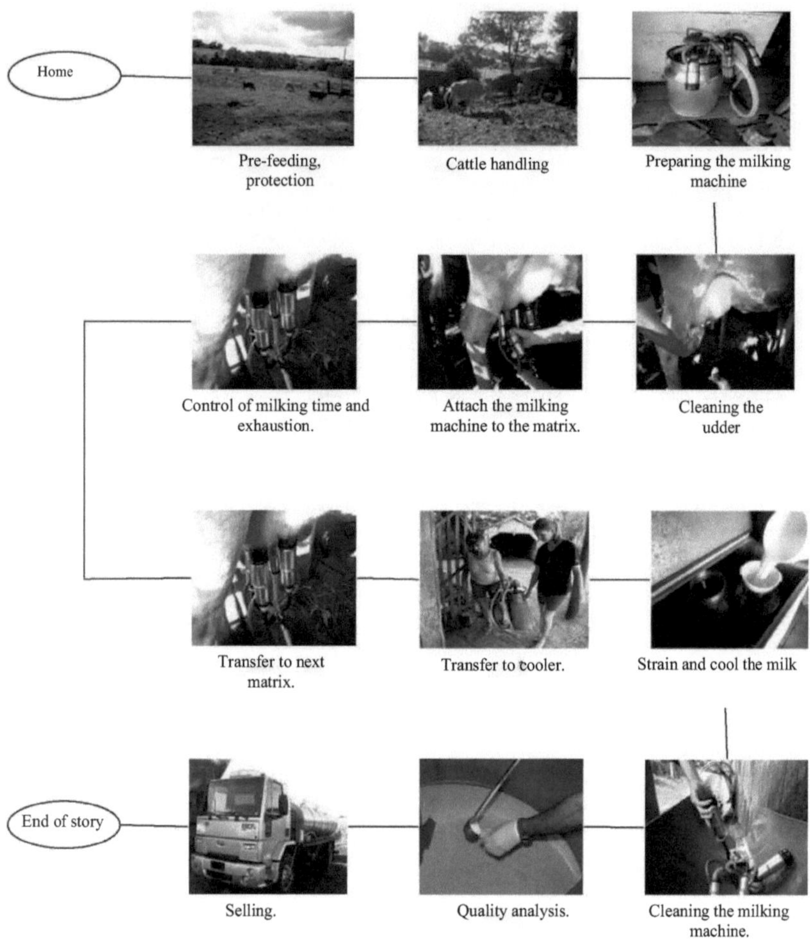

Illustration 4: Milk Production Process Flowchart.
Source: Academic (EBERHARDT,2013)

In order to use the milking machine effectively, it is necessary to assemble and adjust a few items for it to work, adjust the milking machine's liner rubbers, arrange the connecting hoses from the machine to the liners and to the milk pipe, so that the milk can pass through to the milk pipe. Before milking begins, it is extremely important to clean the hose of the matrix, making it possible to attach the machine, starting the time control and the end of milking. At this stage, the sows are fed at the same time.

When the milking machine runs out, it is transferred to another dam and so on, until the milking of lactating dams is complete.

Once this stage of the process is complete, the milk is transferred to the milk cooler, which is strained and placed in tubes for cooling. After this procedure, the milking machine and all the utensils used in the process are washed.

The milk is marketed every two production days, with the main customer being Tche

Laticínios, which receives the product through a transport company that picks up the product from the property under its responsibility. When the product is picked up, the transporter analyzes the quality of the product.

The resources used in the milk production process according to the flowchart in illustration number 4 and the costs identified for its execution are as follows:

Illustration 5: Resources needed for the milk production process

NECESSARY RESOURCES:
1 Sword Physical: fenced enclosure, pasture, silage
2 Labor: for handling equipment
3 Equipment: milking machine, cooler, pasture chopper
4 Installations: milking structure and cooler
5 Water
6 Electricity
7 Products for cleaning equipment
8 Equipment maintenance products (rubbers, hoses, strainers, etc.)
milk and oil for milking machines)
9 breeding stock: Insemination, medicines, veterinarian

Source: Académica (EBERHARDT, 2013)

The resources involved and mentioned in the previous illustration are responsible for the costs inherent in the process, since it is the development of the necessary activities that generates costs and expenses and not the product itself. This way, knowing the activities makes it possible to identify the corresponding values, such as investment in the dairy farm's infrastructure.

The production process from planting to harvesting and marketing soybeans is presented below.

3.1.2 Soybean Production Process

The soybean production process begins with the definition of the area to be cultivated for soybean production. Once the area has been defined, production planning takes place with decisions on the purchase of inputs, the type of seeds, the planting period and the outsourcing of machinery work.

The following flowchart shows all the steps taken in the activity

Illustration 6: Soybean Production Process Flowchart
Source: Academic (EBERHARDT, 2013)

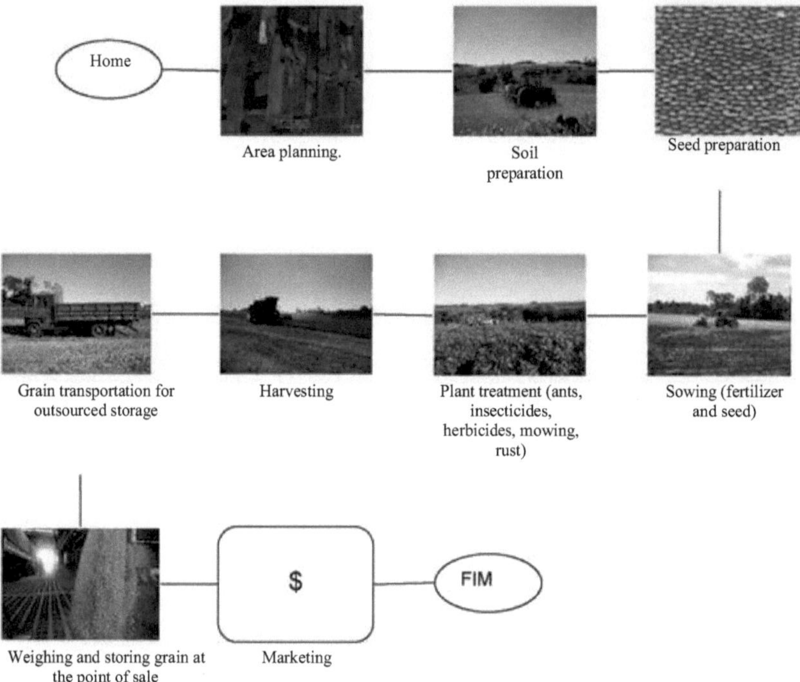

Soil preparation is defined as cleaning, which can be done by plowing, harrowing and/or using fertilizers to keep the soil clean and ready for planting.

Before sowing, it is necessary to treat the seed, which provides greater certainty that the seed is more likely to germinate.

Sowing the seeds and fertilizer takes place from October to December, depending on the weather conditions Soybean treatment is done to protect the plant from pests that negatively affect production, such as insects, caterpillars, rust and plants, and treatment is also done to increase production, such as the use of leaf fertilizer.

After the soybean treatment period, when the plant and grain are dry, the harvest takes place from March to April. Harvesting and transportation of the soya to the point of storage and sale is outsourced. Selling takes place when the owner is in need of funds, or if the time is right to sell.

The resources used in the soybean production process, as shown in illustration number 7, are identified in the following table.

Illustration 7: Resources needed for the
soy production process

RESOURCES NEEDED
Physical space: land
2 Labor: planting, fertilizing, production control, harvesting and marketing
3 Inputs: seeds, manure, fertilizers, pesticides.
4 Equipment (rental): tractor, seeder, sprayer, harvester and truck.
5 Energy (fuel)

Source: Académica (EBERHARDT, 2013)

Having identified the resources involved and mentioned in the previous illustration, the next step is to calculate the cost of each activity or equipment rental to complement the permanent investments. In this process, we can highlight the outlay corresponding to the purchase of quality seeds and their complementary treatment.

In order to continue gathering data to evaluate the results of the production processes under study, the following are spreadsheets with detailed information, with the aim of classifying expenses according to their variability.

3.2 CONTROL OF PROCESS COSTS, BASED ON A SPECIFIC METHODOLOGY.

After identifying the resources involved in each production process on the property that is the subject of this study, the next step is to control the values of each activity carried out, based on surveys of the property's documents and records.

These documents consist of invoices, receipts, contracts and informal control spreadsheets generated by the owner throughout the research period.

The cost of the pasture, the first cost of milk production, according to Appendix A, is calculated by adding up all the inputs used for planting and also the inputs used during grazing, which takes place to conserve the pasture. This data gives the total cost spent on winter and summer pastures.

To calculate the monthly cost, you need to measure the number of months that the pasture will be used to feed the sows. Once you know the number of months, divide the total cost by the months of consumption to get the monthly cost.

Appendix A shows the data collected to calculate pasture costs, which is the first variable cost included in the cost list, as shown in the illustration below, which shows all the costs incurred in the milk production process.

List of milk production process costs from May/2012 to April/2013

Variable costs	May/12	Jun/12	Jul/12	Aug/12	Sep/12	Oct/12	Nov/12	Dec/12	Jan/13	Feb/13	Mar/13	Apr/13
Pastures	R$ 204,00	R$ 204,00	R$ 204,00	R$ 204,00	R$ 204,00	R$ 204,00	R$ 111,25	R$ 111,25	R$ 111,25	R$ 111,25	R$ 111,25	R$ 111,25
Silage	R$ 159,17	R$ 159,17	R$ 122,50	R$ 122,50	R$ 122,50	R$ 122,50	R$ 122,50	R$ 122,50	R$ 175,34	R$ 175,34	R$ 175,34	R$ 175,34
Electricity	R$ 55,84	R$ 61,50	R$ 62,81	R$ 64,45	R$ 59,35	R$ 50,86	R$ 51,08	R$ 50,84	R$ 49,62	R$ 49,98	R$ 49,72	R$ 58,26
Salt	R$ 7,15	R$ 7,15	R$ 7,15	R$ 5,75	R$ 5,75	R$ 5,75	R$ 5,75	R$ 7,98	R$ 7,98	R$ 7,98	R$ 8,65	R$ 8,65
Quirera	R$ 61,81	R$ 61,81	R$ 61,81	R$ 61,81	R$ 61,81	R$ 61,81	R$ 61,81	R$ 61,81	R$ 61,81	R$ 61,81	R$ 61,81	R$ 61,81
Maintenance	R$ 2,96	R$ 9,80										
Milking oil	R$ 3,22	R$ 3,22	R$ 4,83	R$ 4,83	R$ 3,27	R$ 3,27	R$ 3,27	R$ 4,91	R$ 4,91	R$ 5,00	R$ 5,00	R$ 5,00

Insemination				R$ 27,00				R$ 27,00	R$ 62,00	R$ 29,00	R$ 17,00	
Medicines	R$ 40,41					R$ 68,05						
Veterinarian	R$ 80,00					R$ 163,95						
Prod. Cleaning	R$ 4,00	R$ 4,00	R$ 4,00	R$ 4,00	R$ 4,00	R$ 4,00	R$ 4,00	R$ 4,00	R$ 4,00	R$ 4,00	R$ 4,00	R$ 4,00
Water												
Fuel												
Fixed Costs												
Labor												
Milking machine depreciation	R$ 10,00	R$ 10,00	R$ 10,00	R$ 10,00	R$ 10,00	R$ 10,00	R$ 10,00	R$ 10,00	R$ 10,00	R$ 10,00	R$ 10,00	R$ 10,00

Chiller depreciation												
Deprec. Structures	R$ 13,33	R$ 13,33	R$ 13,33	R$ 13,33	R$ 13,33	R$ 13,33	R$ 13,33	R$ 13,33	R$ 13,33	R$ 13,33	R$ 13,33	R$ 13,33
Depreciation	R$ 4,00	R$ 4,00	R$ 4,00	R$ 4,00	R$ 4,00	R$ 4,00	R$ 4,00	R$ 4,00	R$ 4,00	R$ 4,00	R$ 4,00	R$ 4,00
Deprec. Chopper	R$ 1,18	R$ 1,18	R$ 1,18	R$ 1,18	R$ 1,18	R$ 1,18	R$ 1,18	R$ 1,18	R$ 1,18	R$ 1,18	R$ 1,18	R$ 1,18
Deprec. Crusher	R$ 5,28	R$ 5,28	R$ 5,28	R$ 5,28	R$ 5,28	R$ 5,28	R$ 5,28	R$ 5,28	R$ 5,28	R$ 5,28	R$ 5,28	R$ 5,28
Depreciated Matrices	R$ 85,00	R$ 85,00	R$ 85,00	R$ 85,00	R$ 85,00	R$ 85,00	R$ 85,00	R$ 85,00	R$ 85,00	R$ 85,00	R$ 85,00	R$ 85,00
Total Costs	**R$ 737,35**	**R$ 629,44**	**R$ 585,89**	**R$ 613,13**	**R$ 579,47**	**R$ 802,98**	**R$ 478,45**	**R$ 509,08**	**R$ 595,70**	**R$ 563,15**	**R$ 551,56**	**R$ 543,10**

Illustration 8: Spreadsheet of the costs of the milk production process from May 2012 to April 2013

Source: Académica (EBERHARDT,2013)

The cost of silage, the second component of the cost table, is calculated according to Appendix B. Costs are incurred from planting the corn to producing the silage.

The total cost is calculated by adding the cost of planting corn to the cost of producing silage. To obtain the monthly cost, it is necessary to know the months of consumption and divide the total cost by the number of months to obtain the monthly cost.

Electricity, according to Appendix C, is calculated based on the information contained in the motors of the equipment used in milk production. The KW/h were calculated and multiplied by the average cost of the KW charged on the energy bills, in order to obtain the consumption of each piece of equipment. The monthly cost of electricity involves the milking machine, cooler, fodder chopper, crusher and a light bulb.

According to index D, salt is calculated by taking the purchase price and dividing it by the months of consumption. Two types of salt are used to feed the breeders: common salt and mineral salt, and to obtain the monthly cost of salt, the salt costs for each period are added together.

The cost of the quirera is calculated from the production of the corn to its harvest and storage in the silo. The total cost is divided by the number of bags harvested, and the monthly cost of the quirera used to feed the sows is calculated according to the number of bags of corn used per month to produce the quirera, the calculation of which can be seen in Appendix E.

Maintenance is a variable cost that occurs when there is a need to change the handles and components used in milk production. In illustration 8, the maintenance costs refer to changing the sieve of the milk strainer and the short hose used in the milking machine.

The cost of insemination is variable and corresponds to all the expenses incurred by the hired professional, and insemination occurs occasionally when the farm needs genetic improvement.

Medicines and the veterinarian are the costs that arise when the breeding stock fall ill or need some supplement for production, involving the veterinarian's travel and service costs and the cost of the medicines used in the process.

Cleaning products are used to clean and sanitize the milking machine and other equipment used in dairy production. They are calculated according to the quantity consumed during the month, and the consumption of these products is not significant.

The water is taken from a slope, which has the structure of a well, located on the neighboring property, generating no costs for the property.

The fuel used in dairy production is included with the rental of machinery, when corn and pasture are planted.

The cost of labor in dairy farming was not allocated due to the owners having other activities on the property.

The fixed costs represented by depreciation were calculated for all the equipment, structures and sows responsible for dairy production. The methodology applied was based on the value of the asset minus its residual value and divided by the useful life of each production unit.

In the same way, the production costs of soybeans, shown in Appendix H corresponding to the 2011/2012 and 2012/2013 harvests, are obtained.

The costs of the soybean production process start with the purchase of inputs for planting and end with the harvest. The costs of the initial phase involve the purchase of inputs for planting, which are seeds and fertilizer, the cost of planting, which involves renting the tractor and seeder used for planting, as well as the operator's labor included in this value.

After planting, when the soybeans are at a certain level of growth, treatments begin to protect the plant against pests, using pesticides, herbicides, depending on the need for protection and care of the crop. To treat the soybeans, it is also necessary to pay the rent for a sprayer and a tractor to carry out the treatment on the plantation.

When the soybeans are dry, the harvest takes place. Harvesting requires renting a harvester and a truck to transport the soybeans to the warehouse for sale.

When the crop is harvested, a percentage of the total soya harvested is paid as machine rent. In the two periods analyzed, the drivers charged 12% of the harvest actually carried out.

3.3 PROFITABILITY ANALYSIS OF PRODUCTION PROCESSES

Based on the information on the costs of each production process, it became possible to start analyzing the profitability of the property. These procedures involve obtaining information on the income from the process, as well as the relevant expenses.

The first step is to analyze the milk production process, where revenue corresponds to the sale of fresh milk to the industry, in accordance with a previously stipulated agreement. The table below shows revenue, expenses, costs and results for the period from May 2012 to April 2013.

Milk production results

Month	Quantity lt.	Unit price	Gross Revenue	Taxes	Gross Profit	Production costs	Net profit	Net margin
May/12	1880	R$ 0,61	R$ 1.146,80	R$ 26,38	R$ 1.120,42	R$737 ,35	R$ 383,07	33%
Jun/12	1781	R$ 0,60	R$ 1.068,60	R$ 24,58	R$ 1.044,02	R$629 ,44	R$ 414,58	39%
Jul/12	2360	R$ 0,66	R$ 1.557,60	R$ 36,77	R$ 1.520,83	R$585 ,89	R$ 934,94	60%
Aug/12	2074	R$ 0,65	R$ 1.348,10	R$ 31,82	R$ 1.316,28	R$613 ,13	R$ 703,15	52%
Sep/12	1526	R$ 0,65	R$ 991,90	R$ 23,41	R$ 968,49	R$579 ,47	R$ 389,02	39%
Oct/12	1547	R$ 0,67	R$ 1.036,49	R$ 24,47	R$ 1.012,02	R$802 ,98	R$ 209,04	20%
Nov/12	1909	R$ 0,69	R$ 1.317,21	R$ 31,09	R$ 1.286,12	R$478 ,45	R$ 807,67	61%
Dec/12	2260	R$ 0,69	R$ 1.559,40	R$ 36,81	R$ 1.522,59	R$509 ,08	R$ 1.013,51	65%

Jan/13	2045	R$ 0,69	R$ 1.411,05	R$ 33,31	R$ 1.377,74	R$595 ,70	R$ 782,04	55%
Feb/13	1725	R$ 0,69	R$ 1.190,25	R$ 28,10	R$ 1.162,15	R$563 ,15	R$ 599,00	50%
Mar/13	1739	R$ 0,70	R$ 1.217,30	R$ 28,73	R$ 1.188,57	R$551 ,56	R$ 637,01	52%
Apr/13	1530	R$ 0,73	R$ 1.116,90	R$ 26,36	R$ 1.090,54	R$543 ,10	R$ 547,44	49%
Total	**22376,00**	**R$ 0,67**	**R$ 14.961,60**	**R$ 351,83**	**R$ 14.609,77**	**R$7 .189,30**	**R$ 7.420,47**	**50%**

Illustration 9: Results of milk production
Source: Academic (EBERHARDT, 2013)

Gross revenue is obtained by multiplying the unit price with the number of liters sold in each period. The unit price is obtained from the invoice, as shown in Annex B. The quantity is also shown on the invoice and its source is the control generated during daily collection, which is shown in Annex C. The tax burden corresponds to Funrural and Fundesa, also available on the invoices, and the calculation percentages are 2.3% and 0.06% respectively. The calculation basis refers to the total value of the invoice.

Gross profit corresponds to the difference between gross revenue minus the tax burden, and net profit is the difference between gross profit and the cost of production for the period covered by the research, corresponding to twelve months. Based on the same table, it is possible to obtain the average cost of producing a liter of milk, which is R$0.3212, and the average tax expense is R$0.0157, with an average unit result of R$0.3316.

The next section of the study presents data on the soybean production process, including the number of bags harvested and the corresponding costs.

The table below shows the market value, costs, expenses and income from the sale of soybeans.

Yields from the soybean production process						
Mes	**May/12**	**18/08/2012**	**20/08/2012**	**21/08/2012**	**07/12/2012**	**Total**
Quantity Kg	1860	2400	600	2090	250	7200

Quantity of 60 kg bags	31	40	10	34,83	4,17	120
Unit price Kg	R$0 ,88	R$0 ,75	R$ 0,65	R$ 0,67	R$ 1,00	R$ 0,79
Nail sacks	R$53 ,00	R$ 45,00	R$ 39,00	R$ 40,20	R$ 60,00	R$ 47,44
Sales price	R$ 1.643,00	R$ 1.800,00	R$ 390,00	R$ 1.393,00	R$ 250,00	R$ 5.476,00
Taxes	R$37 ,79	R$ 63,48	R$ 15,87	R$ 56,08	R$ 5,75	R$ 178,97
Gross Profit	R$ 1.605,21	R$ 1.736,52	R$ 374,13	R$ 1.336,92	R$ 244,25	R$ 5.297,03
Unit cost	R$0 ,40	R$0 ,40	R$ 0,40	R$ 0,40	R$ 0,40	R$ 0,40
Total Cost	R$ 749,58	R$ 967,20	R$ 241,80	R$ 842,27	R$ 100,75	R$ 2.901,60
Net profit	**R$ 855,63**	**R$ 769,32**	**R$ 132,33**	**R$ 494,65**	**R$ 143,50**	**R$ 2.395,43**
Net margin	**52%**	**43%**	**34%**	**36%**	**57%**	**44%**

Illustration 10: Result of the soybean production process
Source: Académica (EBERHARDT, 2013)

The cost and sales figures for soybeans shown in the table above refer to the soybean production process for the 2011/2012 harvest. The months shown in the table correspond to the months in which the bags were sold. The costs of this production correspond to the period from planting to harvest, from November 2011 to April 2012.

The selling price of a sack of soybeans is set by the market and fluctuates over short periods of time. The taxes corresponding to the sale of soybeans are related to Funrural, which is 2.3%, and the rest of the discount is related to *royalties* charged by transgenics, on the total value of the sale. Only the sale that took place in December does not have a *royalty* discount, only Funrural. The difference between the sale price and the tax is the gross profit of the operation.

The costs are related to the calculation of soybean production costs available in Appendix H. Net profit is the difference between gross profit and the total cost of each trade made.

It can be seen that the net profit from marketing soybeans is directly related to the price set by the market. The higher the selling price, the higher the net profit, because the total cost of each bag is the same during the marketing period of a given harvest.

The results show an average profit per bag of R$19.96 and an average cost of R$24.18. The average sale price in the period was R$47.44 and the average turnover was R$45.63 per bag of soy sold.

During this period, the harvest was considered below average, increasing the unit cost of production. The average production was 30 bags per hectare, due to drought at the time of plant development, causing low productivity.

Another factor that reduced the result was marketing after the May period, when sales prices were reduced, as shown in illustration 10.

3.3.1 Profitability of Milk and Soybean Production

With the results on the profitability of the milk production process and the soybean harvest for the periods determined for this study, it became possible to identify the profitability of the businesses.

It is necessary to obtain additional data for the calculation procedures, such as the amounts of investments in fixed assets and working assets.

The investment in milk production is primarily in the amount of land needed to carry out the production process, especially the production of feed for the dairy cattle.

The equipment and structure used are considered investments in the process, and as a basis for calculation their values are already discounted for depreciation.

Milk production investment			
Investments	**Quantity**	**Unit value**	**Total**
Land hectare	6	R$ 20.000,00	R$ 120.000,00
Structures	1	R$1.960,00	R$ 1.960,00
Fenced enclosure	1	R$ 588,00	R$ 588,00
Milking machine	1	R$ 120,00	R$ 120,00
Cooler	1	-	-

Fodder chopper	1	R$	50,60	R$	50,60
Shredder/grinder	1	R$	103,33	R$	103,33
Working capital	1	R$3.241,04		R$	3.241,04
Matrices	8	R$1.431,00		R$11.450,00	
Equipment replacement	2			R$	5000,00
Total investment				**R$142.512,97**	

Table 1: Investment in the milk production process
Source: Académica (EBERHARDT, 2013)

Working capital corresponds to the costs of the last six months of milk production. It refers to the resources needed to maintain the milk production process.

And the matrices, which are mainly responsible for the production process, have their value stipulated by the price sold on the local market and the net value of depreciation. The last item indicates the need to purchase some equipment that is already fully depreciated and in a precarious state.

With the total value of the investment made and identified in Table 1, it is possible to calculate the profitability of the process, which is the return on the investment made in the milk process.

Profitability of the milk production process	
Period	May/12 to Apr/13
Investment	R$ 142.512,97
Total Net Profit	R$7 .420,47
Annual profitability	5,21%

Table 2: Profitability of the milk production process
Source: Académica (EBERHARDT, 2013)

The calculation shown in Table 2 indicates a return of 5.21% **p**.a., which means that for every R$100.00 invested, the farmer earns R$5.21. This return corresponds to a 12-month period and is similar to the return offered by low-risk financial investments. This profitability, in rural activities, is considered remuneration for the work done by the rural producer, and in order to be considered adequate or not, it must be compared with a benchmark compatible with its characteristics and particularities.

In monetary terms, this profitability corresponds to an average wage of R$618.37 for two people, with an average dedication of 3 hours a day. This partial dedication indicates that other activities take place on the property and complement the producers' income. This study presents an analysis of the profitability of soybean production based on the 2011/2012 and 2012/2013 harvests.

In the same way, the procedures for analyzing the soybean harvest involve a survey of the investments made in order to plant, develop the grains and harvest them.

For the soybean production process, the investments shown in the table below were used:

Safra	Investments made in the soy production process			
2011/2012	Investments	Quantity	Unit value	Total
	Land hectare	4	R$10　　　.000,00	R$ 40.000,00
	Working capital	1	R$2　　　.902,13	R$ 2.902,13
	Total			**R$ 42.902,13**
2012/2013				
	Land hectare	4	R$10　　　.000,00	R$ 40.000,00
	Working capital	1	R$3　　　.936,03	R$ 3.936,03

	Total			R$ 43.936,03

Table 3: Investment in soybean production Source: Academic (EBERHARDT, 2013)

In order to carry out the soybean production process, a working capital of R$3,936.03 is needed for each harvest, which corresponds to the current cost of production for the 2012/2013 harvest. The amount of working capital corresponds to four hectares of soybean cultivation.

The value of the land represents only half of its value, because the soybean harvest is only for a period of six months.

With the total value of the investment made in the soybean production process, it is possible to check how much the product is bringing in for the property, for its investors.

The table below calculates the profitability of the soybean production process based on the 2011/2012 harvest.

Profitability of the soy production process		
Period	2011/2012 harvest	2012/2013 harvest
Investment	R$ 42.902,13	R$ 43.813,63
Total Net Profit	R$ 2.395,43	R$ 6.029,37
Crop profitability	5,58%	13,76%

Table 4: Profitability of the soybean production process Source: Academic (EBERHARDT, 2013)

For the 2012/2013 harvest, since there was no trading during the period analyzed, the price stipulated by the market on 14/06/13 was used as a reference, with a value of R$60.00 per bag of soybeans, as shown in Annex F.

The profitability of soy corresponds to the owner's remuneration. In monetary terms, the profitability of soybeans from the 2011/2012 harvest corresponds to R$199.61 per month, which means that for every R$100.00 invested, the owner gets a return of R$5.58, which is added to the owner's total remuneration.

In relation to the 2012/2013 harvest, the monthly profitability figures are R$502.44, which means that for every R$100.00 invested, the owner gets a return of R$13.76. The last harvest showed a significant increase in results compared to the previous harvest, due to the weather conditions and the more favorable market price in the current period.

For this reason, the return on agricultural production depends not only on the efforts made by human resources and the application of inputs, but also on the compatibility of climatic conditions and market prices for a good return on production.

3.4 ANALYSIS OF BUSINESS ALTERNATIVES FOR THE RURAL PROPERTY UNDER STUDY

The rural property that is the subject of this study has two main productions, soy and milk, so it needs to know the best production alternative that will bring the highest return.

As an alternative, it offers leasing, which is renting the land to third parties who plant and harvest it themselves and at the end of the harvest, the lessor is paid a percentage of the production. The current rent paid in the region for soya is 30% of the total amount harvested. For the winter crop, wheat, the rent is 10% of the total amount harvested.

The following table shows a comparison of production and yields in relation to the soybean crop:

Lease 4 hectares of soya production				
Soy	Quantity produced	Production Profit	Quantity Rent	Renting
2011/2012 harvest	120	R$2.395,43	36	R$1.811,00
2012/2013 harvest	170	R$6.029,37	51	R$2.640,83
Total	290	R$8.424,80	87	R$4.451,83

Table 5: Comparison of soybean production and leasing Source: Academic (EBERHARDT, 2013)

The rental value of the soybean crops was calculated on the soybean sales price in April 2012 and April 2013 respectively, according to appendix G and H, the period in which the soybean is harvested and payment is made to the landowner.

With soybean production, the owners have a monthly return of R$199.61 for the 2011/2012 harvest and R$502.44 for the 2012/2013 harvest.

With the option of leasing the land to produce soybeans, the landowner would have a monthly remuneration, corresponding to the 2011/2012 harvest, of R$150.91 and for the 2012/2013 harvest, a monthly monetary value corresponding to R$220.06.

Compared to the two alternatives, it can be seen that producing soybeans on the property brings more return, remuneration to the owner, with the rental value representing half the sum of the profit over the period analyzed, a lower value than the remuneration that occurs if the owner continues to produce soybeans. With production, the landowner guarantees more pay for his workforce and more return on his investment.

In the same way, the leasing of milk production was analyzed by comparing the results of leasing the area to soybeans and wheat.

Milk lease 6 hectares May/12 to Apr/13				
		Annual Profit		Monthly Profit
Milk Production	R$	**7.420,47**	R$	**618,37**
Soy leasing	R$	3.961,25	R$	337,87
Renting Wheat	R$	923,26	R$	76,94
Total rent	R$	**4884,50**	R$	**407,04**

Table 6: Comparison of milk production and rent Source: Académica (EBERHARDT, 2013)

The value of the land leased for planting soybeans is proportional to the 2012/2013 harvest, with the sale value stipulated in the month of the harvest, according to Annex H, where the agreed value is paid or transferred to the rent, which in this case would be 30% of the total number of bags harvested. The annual profit is discounted by Funrural.

The value of the wheat rent corresponds to 10% of the quantity harvested. The average yield of the wheat crop is 50 bags per hectare, and the wheat price was used at harvest time, according to Annex E. Using the average price from Annex E, it was possible to identify the profit already deducted from Funrural.

By analyzing this table, it is possible to identify that it is better for the rural property to produce than to lease the land to grow soybeans and wheat, since milk production has a much higher profit than the sum of leasing the land with soybeans and wheat in the same period as milk production.

However, as well as milk and soya being the farm's main productive activities, it also has other subsistence activities, which also bring returns to the property under study.

Chapter 4

4 RECOMMENDATIONS

With the analysis of the rural property's production processes, identifying the costs of each process, their results, profitability and return, it is recommended that the rural property continue with the production processes as they bring more return to the property than the alternative business in comparison.

As a result, results can change. Soybean production can change due to various factors such as weather conditions, market conditions of supply and demand, high or low crop yields in other countries and producing regions.

Milk can change its results with the emergence of new buyers, as well as genetic improvement with increased productivity, and with some reductions in costs, all of which improve the results of the process.

It is therefore important that rural properties continue to maintain internal controls over their production, so that they can evaluate possible future alternatives and new businesses in rural areas.

The rural environment has many business alternatives and many production possibilities, and for this reason it is recommended that the results of current production be evaluated with the alternatives in order to identify the production that will bring the greatest return on the investments made.

More studies are therefore recommended in the rural area, as it is an area undergoing major productive expansion and needs studies and analysis to manage and help make decisions on rural properties.

CONCLUSION

The current scenario of rural properties requires technological evolution and management tools so that their processes are well planned with competitive capacity and can maximize results.

In this study, describing the production processes made it possible to identify the costs of each milk and soy production process. And by identifying the costs of the processes, it was possible to analyze and have a position on the profits of each production.

As for determining the methodology used to control and record expenses, it was determined to be hybrid, partly because it is more relevant to variable costs. Variable costs are combined with fixed costs to arrive at the total cost of milk production. However, soybean production has no fixed costs, as the property rents the machinery and equipment for the production process. It only has variable costs, which represent the total production costs.

By identifying the costs of each production process and the investments made, it became possible to assess the profitability of milk and soybean production. The profitability of dairy production is 5.21%, representing a return equivalent to low-risk investments, considered a stable return for the value of the investment.

The profitability of soybean production is totally related to the selling price set by the market and the weather conditions, which are mainly responsible for the productivity of each harvest. In the 2011/2012 harvest, the return was 5.58%, due to the drought which caused low productivity in this harvest, and the sale of soybean bags at a time when the price was not very favorable. In the 2012/2013 harvest, profitability increased significantly to 13.76% on the crop's investment. The increase in profitability in the last harvest was due to the good price paid by the market and the increase in productivity due to the drought having occurred for a shorter period this year.

It can be seen that soybean production in the last harvest was almost on a par with dairy farming, as it has been for the last twelve months. With this analysis, in six months soya has brought in almost the same income as dairy farming.

Leasing the soybean crop would bring less return to the property than production itself. With soybean production, the property has a good return compared to renting, and this return has already been deducted from all production costs.

Similarly, when analyzing the alternatives of soybean and wheat production to dairy farming, it can be seen that the sum of the profits from the two leases was not equal to the profits from dairy farming, with dairy farming being the most profitable in the period under study.

The minimum result desired by the entrepreneur when analyzing the alternatives is 30% for soybean production and 10% for leasing wheat production. These minimum desired results are easily achieved with milk and soybean production, with average profitability of 50% and 44% respectively.

It is important for rural properties to have internal controls that help them make decisions about their production and results, thus obtaining reliable decisions that meet the expectations of the minimum return desired by the rural owners and for the investments made.

Therefore, at the end of this study, based on the bibliographical research for the realization of the research, it is considered that the study has added knowledge and professional practice. It has also added knowledge and assessment of controls on rural property, helping with future decisions on the property.

REFERENCES

ALMEIDA, Kátia de; MACEDO, Marcelo Álvaro da Silva. **Analysis of Accounting and Financial Performance in Brazilian Agribusiness: Applying DEA to the Agribusiness Sector in 2006 and 2007**. Revista Pensar Contábil, Rio de Janeiro, v.12, n.48, p.5-21, mai/ago.2010.

ALMEIDA, Marcelo Cavalcanti. **Internal Auditing**. A Modern and Complete Course. 6ª ed., Sao Paulo: Atlas, 2008.

ANDRIOLI, Antônio Inácio. **Family Farming and Environmental Sustainability**. Espaço Académico Magazine, No. 89, October 2008. Available at: <http://www.espacoacademico.com.br/089/89andrioli.pdf> Accessed on: 13/12/2012

ARAÚJO, José Lincoln Pinheiro; ARAÚJO, Edílson Pinheiro; CORREIA, Rebert Coelho. **Analysis of production costs and profitability of the onion crop in the Säo Francisco sub-medium region.** XV SIMPEP Symposium on Production Engineering, November 10-12, 2008, Information Systems and Knowledge Management. Available at: <http://www.alice.cnptia.embrapa.br/bitstream/doc/15817/1/OPB2093.pdf> Accessed on: 06/11/2012

ASSAF NETO, Alexandre. **Corporate Finance and Value**. 5th ed., Sao Paulo: Atlas, 2010.

. **Structure and Analysis of Balance Sheets**. 8th edition, Sao Paulo: Atlas, 2009.

ATKINSON, Anthony A. et al. **Management Accounting**. 3rd ed., Sao Paulo: Atlas, 2011.

BEUREN, Ilse Maria, et al. **Como elaborar Trabalhos Monográficos Contabilidade, Teoria e Prática**. Sao Paulo: Atlas, 2004.

BRAZIL, Federal Revenue Service. **DEPRECIATION - SRF Normative Instruction No. 162 of December 31, 1998**. Available at: <http://www.receita.fazenda.gov.br/Legislacao/ins/Ant2001/1998/in16298.htm> Accessed on:26/04/2013.

CANZIANI, José Roberto. **The Management of Rural Enterprises**. Rural Entrepreneur Program. 2nd volume. Curitiba: SEBRAE/PR and SENAR/PR, 3 volumes, 2009.

Agroindustrial Chains. Rural Entrepreneur Program. 2nd volume. Curitiba: SEBRAE/PR and SENAR/PR, 3 volumes, 2009.

CARRON, Wilson; GUIMARÁES, Osvaldo. **Physics**. 2ª ed., Sao Paulo: Moderna, 2003.

CERVO, Amado Luiz; BERVIAN, Pedro Alcino; DA SILVA, Roberto. **Scientific methodology**. 6ª ed., Sao Paulo: Pearson Prentice Hall, 2007.

CONAB, National Supply Company. **Cost of agricultural production, Conab's methodology,** 2010 Available at: <HTTP://www.conab.gov.br/olalaCMS/uploads/arquivos/0086a569bafb14cebf87bd11 1936e115..pdf> Accessed on: 17/08/2012

COTRIROSA, Cooperativa Tríticola Santa Rosa**, October 2012 wheat price quotation**. Available at: <http//:www.cotrirosa.com.br>. Accessed on: 14/06/2013.

Price of a sack of soybeans in June 2013. Available at: <http//:www.cotrirosa.com.br>. Accessed on: 14/06/2013.

. **Price of a sack of soybeans in April 2012.** Available at:<http//:www.cotrirosa.com.br> . Accessed on: 14/06/2013.

. **Price of a sack of soybeans in April 2013.** Available at: <http//:www.cotrirosa.com.br>. Accessed on: 17/06/2013.

CREPALDI, Sílvio Aparecido. **Management Accounting**. 6th ed., Sao Paulo: Atlas, 2012.

. **Rural accounting: a decision-making approach**. 5th ed., Sao Paulo: Atlas, 2009.

CRUZ, Tadeu. **Systems, Organization and Methods**. 3rd ed., Sao Paulo: Atlas, 2011.

EARTH, Google. **Photo of rural property**. Available at: <http//:www.earth.google.com>. Accessed on 18/08/2012.

FURASTÉ, Pedro Augusto. **Technical norms for scientific work - preparation and formatting. Explanation of the ANBT Standards**. 14th ed., Porto Alegre: s.n., 2008.

GIL, Antonio Carlos. **How to design research projects**. 5th ed., Sao Paulo: Atlas, 2010.

GITMAN, Lawrence J. **Principles of Financial Management**. 12th ed., Sao Paulo: Pearson Prentice Hall, 2010.

HENDRIKSEN, Eldon S.; VAN BREDA, Michael F. **Teoria da Contabilidade**. 1st ed., 10th reprint, Sao Paulo: Atlas, 2012.

HORNGREN, Charles T.; SUNDEN, Gary L.; STRATTON, Willian O. **Management Accounting**. Translated into Portuguese by Elias Pereira. 12ª ed., Sao Paulo: Prentice Hall, 2004.

IUDÍCIBUS, Sérgio de. **Management accounting**. 6th ed., Sao Paulo: Atlas S.A, 2009.

MARCONI, Marina de Andrade; LAKATOS, Eva Maria. **Fundamentals of scientific methodology**. 7th ed., Sao Paulo: Atlas 2010.

MARION, José Carlos. **Rural accounting:** agricultural accounting, livestock accounting, corporate income tax. 11th ed., Sao Paulo: Atlas, 2010.

MARION, José Carlos; SEGATTI, Sónia. **Cost management system in small dairy farms.** Custos e Agronegócios *online* - v.2 - n.2-Jul/Dec - 2006 ISSN 1808-2882. Available at: <http//:www.custoseagronegocioonline.com.br> Accessed on: 25/08/2012.

MARTINS, Eliseu. **Cost Accounting**. 9th ed., Sao Paulo: Atlas, 2008.

MEDEIROS, Joao Bosco. **Redagão CientíficaA prática de Fichamentos, Resumos, Resenhas**. 11th ed., Sao Paulo: Atlas, 2010.

MEGLIORINI, Evandir. **Costs: Analysis and Management.** 3rd ed., Sao Paulo: Pearson Prentice Hall, 2012.

MOTTA, Regis da Rocha; CALOBA, Guilherme Marques. **Investment Analysis**: Decision-making in Industrial Projects. Sao Paulo: Atlas, 2009.

NBC T 19.41. **Accounting for small and medium-sized enterprises**. Regional Council of Rio Grande do Sul, Porto Alegre, 2011.

NEUKIRCHEN, Leandro César; ZANCHET, Aládio; PAULA, Germano de,**Tecnologia de gestão e rentabilidade na pequena propriedade rural - estudo de caso**, in Congresso da Sociedade Brasileira de Economia, Administrado e Sociologia Rural, n°?, 2005,vRibeirao Preto. Anais, Ribeirao Preto: SOBER, 2005. Available at:<http://www.sober.org.br/palestra/2/506.pdf>Accessed on 16/10/2012

NOGUEIRA, Daniel Ramos. **Cost Accounting: Accounting Sciences**. Sao Paulo: Pearson Education do Brasil, 2009.

PADOVEZE, Clóvis Luís. **Management Accounting, A Focus on Accounting Information Systems**. 7th ed. Sao Paulo: Atlas, 2010.

PASQUALINI, Fernanda, LOPES, Alceu de Oliveira, SIEDENBERG, Dieter. **Production Management.** Editora Unijuí, 2010.

PEREIRA, Valdecir de Oliveira. **Cost analysis in the custom manufacturing industry**. 7th Meeting of the Tool, Mold and Die Chain, Sao Paulo 2009. Available at: <http://www.abmbrasil.com.br/seminarios/moldes/2009/>. Accessed on: 12/10/2012.

RIBEIRO, Osni Moura. **Cost accounting**. Sao Paulo: Sarai6va, 2009

RIBEIRO, Otilia Denise Jesus. **Adjusting costs in agricultural activity**. Revista Eletrónica de Contabilidade Curso de Ciencias Contábeis UFSM, Volume 1, N.1, Sep-Nov/2004

RIZZON, Luiz Antenor. **Embrapa Uva e vinho, Vinegar production system**. 13 ISSN 1678-8761 versao electrónico Dec/2006. Available at: <http://sistemasdeproducao.cnptia.embrapa.br/FontesHTML/Vinagre/SistemaProduc

aoVinagre/custo.htm> Accessed on: 24/08/2012.

SEGALA, Cristiane Zucchi Sopelsa; SILVA, IvanirTechioda. **Costing in milk production on a rural property in the municipality of Irani-SC.** Custos e Agronegócio *online* v.3, n.1 - Jan/Jun - 2007 ISSN 1808-2882, Available at: <www.custoseagronegóciosonline.com.br.> Accessed on:07/09/2012.

SIEDENBERG, Dieter; PASQUALINI, Fernanda. **Management of the production of goods and services**. Editora Unijuí 2009.

SILVA, Antonio Carlos Ribeiro da.**Metodologia** da **Pesquisa Aplicada a Contabilidade**. Sao Paulo, Editora Atlas S.A 2008.

SILVA, Edna Lúcia da; MENEZES, EsteraMuszkat. **Metodologia da pesquisa e elaboração de dissertação**. 3ª ed. revised and updated, Florianópolis: Laboratório de Ensino a Distancia da UFSC, 2001.

SZUSTER, Natan, et al. **Contabilidade Geral Introdugao a contabilidade societária**. 2a ed., Sao Paulo: Atlas, 2008.

UENO, Paulo. **Physics**. 1st ed., Sao Paulo: Ática, 2006.

VALLE, Rogerio; OLIVEIRA, Saulo Barbará de (Org.). **Analysis and Modeling of Business Processes**. Sao Paulo: Atlas, 2012.

VELTER, Francisco; MISSAGIA, Luiz. **Cost Accounting and Analysis of Financial Statements**. Rio de Janeiro: Elsevier, 2010.

VICECONTI, Paulo Eduardo; NEVES, Silvério das;.**Contabilidade de custos: um enfoque direto e objetivo**. 11th ed., Sao Paulo: Saraiva, 2013.

VIEIRA, Eusélia Paveglio. **Cost and Sales Price Formation**. Ijuí, Editora Unijuí, 2008.

YAMAGUCHI, Luis Carlos Tak ao; GOMES, Aloísio Teixeira; CARNEIRO, Alziro Vasconcelos. **How to calculate profitability in dairy farming**. Embrapa Gado de leite. 2nd edition revised and updated in March 2006. ISSN N° 1518-3254

APPENDICES
APPENDIX A - WINTER AND SUMMER PASTURE COSTS

Winter grazing - May 2012 to October 2012			
Inputs	**Quantity**	**Total**	**Source**
Fertilizer	12 bags	R$624,00	Supplier with invoice
Oats	10 bags	R$ 240,00	Single receipt
Tractor driver shoveling 1	1	R$40,00	Rental
Tractor driver sowing	1	R$160,00	Rental
Manure	4 loads	R$ 160,00	Single receipt
Total		R$1.224,00	
Total Cost	**Months**	**Monthly Cost**	
R$ 1.224,00	6	R$ 204,00	Total costs / 6 months

Summer pasture Period November 2012 to April 2013			
Inputs	Quantity	Total	Source
Fertilizer	4 bags	R$ 245,00	Supplier with invoice
Oats millet	32 Kg	R$ 119,50	Supplier with invoice
Sorghum	2Kg	R$13,00	Supplier with invoice
Tractor driver shoveling	2	R$85,00	Rental
Tractor driver 2 patear	1	R$35,00	Rental
Grid tractor driver	1	R$20,00	Rental
Tractor driver 2 sowing	1	R$70,00	Rental
Tractor driver Planting	1	R$80,00	Rental

Total		R$667,50	

Total Cost	**Months**	**Monthly Cost**	
R$ 667,50	6	R$ 111,25	Total costs/ 6 months

APPENDIX B - COST OF SILAGE

Silage 1st semester 2012			
Inputs	**Quantity**	**Total**	**Source**
Plantation			
Poison	1	R$ 30,00	Supplier with invoice
Poison tractor driver	1	R$ 25,00	Rental
Seed	1 bag	R$ 60,00	Exchange system
Fertilizer	5 bags	R$300,00	Supplier with invoice
Tractors	1	R$ 50,00	Rental
Urea	1 bag	R$ 50,00	Supplier with invoice

Silage			
Tractor driver	1	R$250,00	Rental
tractor driver 2	1	R$ 50,00	Rental
Tractor driver 3	1	R$ 20,00	Rental
Canvas	1	R$ 80,00	Supplier with invoice
Labor	2	R$ 40,00	Single receipt
Total		**R$955,00**	
Total Cost	**Months**	**Monthly Cost**	
R$ 955,00	6	R$ 159,16	Total costs / 6 months

Note: When planting corn, there are the costs of the inputs used to plant and cultivate it. The time to make silage is when the grain is well formed, and its stage is moving from green corn to maturity.

The costs of silage production are spread over the rental of equipment for cutting the corn, loading the crushed corn and unloading it into the silo. It also has some labor costs, which are charged by the hour. The cost of maintaining the silage after it is finished is the purchase of a tarpaulin to cover the silage in the silo, with the aim of preserving the silage without losses occurring.

Silage 2nd semester 2012			
Inputs	**Quantity**	**Total**	**Source**

Plantation			
Seed	1	R$ 67,00	Exchange system
Fertilizer	4	R$208,00	Supplier with invoice
Tractors	1	R$ 50,00	Rental
Urea	2	R$100,00	Supplier with invoice
Silage			
Tractor driver	1	R$150,00	Rental
Tractor driver 2	1	R$ 50,00	Rental
Tractor driver 3	1	R$ 20,00	Rental
Canvas	1	R$ 50,00	Supplier with invoice
Labor	3	R$ 60,00	Single receipt
Total		**R$735,00**	

Total Cost	Months	Monthly Cost	
R$ 735,00	6	R$ 122,50	Total costs/ 6 months

Silage 1st semester 2013			
Inputs	**Quantity**	**Total**	**Source**
Planting corn			
Poison	1	R$ 40,00	Supplier with invoice
Poison tractor driver	1	R$ 20,00	Rental
Seed	1	R$ 70,00	Exchange system
Fertilizer	5	R$315,00	Supplier with invoice
Tractors	1	R$ 60,00	Rental
Urea	1,5	R$ 93,00	Supplier with invoice
Silage			
Silage tractor driver	1	R$250,00	Rental

Tractor driver 2	1	R$ 50,00	Rental
Tractor driver 3	1	R$ 30,00	Rental
Canvas	1	R$ 84,04	Supplier with invoice
Labor	2	R$ 40,00	Single receipt
Total		**R$1.052,04**	
Total Cost	**Months**	**Monthly Cost**	
R$ 1052,00	6	R$ 175,33	Total costs/ 6 months

APPENDIX C - COST OF ELECTRICITY

Cost Electricity Milking machine					
Month	**Kw/h**	**Monthly hours**	**Total kw/h**	**Kw/h value**	**Monthly cost**
May/12	1,87	40	74,8	R$ 0,28	R$20,60
Jun/12	1,87	40	74,8	R$0,30	R$22,68
Jul/12	1,87	40	74,8	R$0,31	R$23,16

Aug/12	1,87	40	74,8	R$0,32	R$23,77
Sep/12	1,87	40	74,8	R$ 0,31	R$23,48
Oct/12	1,87	40	74,8	R$0,31	R$23,55
Nov/12	1,87	40	74,8	R$0,32	R$23,65
Dec/12	1,87	40	74,8	R$0,31	R$23,54
Jan/13	1,87	40	74,8	R$0,31	R$22,97
Feb/13	1,87	40	74,8	R$0,31	R$23,14
Mar/13	1,87	40	74,8	R$0,31	R$23,02
Apr/13	1,87	40	74,8	R$0,29	R$21,49

Cost ElectricityMilk cooler					
Month	Kw/h	Monthly hours	Total kw/h	Kw/h value	Monthly cost

May/12	0,968	75	72,6	R$0,28	R$19,99
Jun/12	0,968	75	72,6	R$0,30	R$22,01
Jul/12	0,968	75	72,6	R$0,31	R$22,48
Aug/12	0,968	75	72,6	R$0,32	R$23,07
Sep/12	0,968	75	72,6	R$0,31	R$22,79
Oct/12	0,968	75	72,6	R$0,31	R$22,85
Nov/12	0,968	75	72,6	R$0,32	R$22,95
Dec/12	0,968	75	72,6	R$0,31	R$22,84
Jan/13	0,968	75	72,6	R$0,31	R$22,30
Feb/13	0,968	75	72,6	R$0,31	R$22,46

Mar/13	0,968	75	72,6	R$0,31	R$22,34
Apr/13	0,968	75	72,6	R$0,29	R$20,85

Cost Electricity Shredder and chipper					
Months	Kw/h	Monthly hours	Total kw/h	Kw/h value	Monthly cost
May/12	2,75	4,5	12,375	R$0,28	R$3,41
Jun/12	2,75	4,5	12,375	R$0,30	R$3,75
Jul/12	2,75	4,5	12,375	R$0,31	R$3,83
Aug/12	2,75	4,5	12,375	R$0,32	R$3,93
Sep/12	2,75	4,5	12,375	R$0,31	R$3,88
Oct/12	2,75	4,5	12,375	R$0,31	R$3,90
Nov/12	2,75	4,5	12,375	R$0,32	R$3,91

Dec/12	2,75	4,5	12,375	R$0,31	R$3,89
Jan/13	2,75	4,5	12,375	R$0,31	R$3,80
Feb/13	2,75	4,5	12,375	R$0,31	R$3,83
Mar/13	2,75	4,5	12,375	R$0,31	R$3,81
Apr/13	2,75	4,5	12,375	R$0,29	R$3,55

Cost Electricity Fodder chopper					
Mes	Kw/h	Monthly hours	Total kw/h	Kw/h value	Monthly cost
May/12	2,75	15	41,25	R$0,28	R$11,36
Jun/12	2,75	15	41,25	R$0,30	R$12,51
Jul/12	2,75	15	41,25	R$0,31	R$12,77
Aug/12	2,75	15	41,25	R$0,32	R$13,11
Sep/12	2,75	10	27,5	R$0,31	R$8,63

Oct/12	2,75		0	R$0,31	R$-
Nov/12	2,75		0	R$0,32	R$ -
Dec/12	2,75		0	R$0,31	R$ -
Jan/13	2,75		0	R$0,31	R$ -
Feb/13	2,75		0	R$0,31	R$ -
Mar/13	2,75		0	R$0,31	R$ -
Apr/13	2,75	15	41,25	R$0,29	R$11,85

Cost Electricity Lamp

Months	Kw/h	Monthly hours	Total kw/h	Kw/h value	Monthly cost
May/12	0,06	30	1,8	R$ 0,28	R$ 0,50
Jun/12	0,06	30	1,8	R$ 0,30	R$ 0,55
Jul/12	0,06	30	1,8	R$ 0,31	R$ 0,56

Aug/12	0,06	30	1,8	R$ 0,32	R$ 0,57
Sep/12	0,06	30	1,8	R$ 0,31	R$ 0,56
Oct/12	0,06	30	1,8	R$ 0,31	R$ 0,57
Nov/12	0,06	30	1,8	R$ 0,32	R$ 0,57
Dec/12	0,06	30	1,8	R$ 0,31	R$ 0,57
Jan/13	0,06	30	1,8	R$ 0,31	R$ 0,55
Feb/13	0,06	30	1,8	R$ 0,31	R$ 0,56
Mar/13	0,06	30	1,8	R$ 0,31	R$ 0,55
Apr/13	0,06	30	1,8	R$ 0,29	R$ 0,52

Monthly Electricity Cost					
Month	Milking machine	Cooler	Fodder chopper	Crusher	Light bulb	Total

May/12	R$20,60	R$19,99	R$11,36	R$ 3,41	R$ 0,50	R$ 55,84
Jun/12	R$22,68	R$22,01	R$12,51	R$ 3,75	R$ 0,55	R$ 61,50
Jul/12	R$23,16	R$22,48	R$ 12,77	R$ 3,83	R$ 0,56	R$ 62,81
Aug/12	R$23,77	R$23,07	R$ 13,11	R$ 3,93	R$ 0,57	R$ 64,45
Sep/12	R$23,48	R$22,79	R$ 8,63	R$ 3,88	R$ 0,56	R$ 59,35
Oct/12	R$23,55	R$22,85	R$ -	R$ 3,90	R$ 0,57	R$ 50,86
Nov/12	R$23,65	R$22,95	R$ -	R$ 3,91	R$ 0,57	R$ 51,08
Dec/12	R$23,54	R$22,84	R$ -	R$ 3,89	R$ 0,57	R$ 50,84
Jan/13	R$22,97	R$22,30	R$ -	R$ 3,80	R$ 0,55	R$ 49,62
Feb/13	R$23,14	R$22,46	R$ -	R$ 3,83	R$ 0,56	R$ 49,98

Mar/13	R$23,02	R$22,34	R$ -	R$ 3,81	R$ 0,55	R$ 49,72
Apr/13	R$21,49	R$20,85	R$ 11,85	R$ 3,55	R$ 0,52	R$58,26

The values for calculating the electrical energy used in the milk production process were based on the values contained in the information sheet on the electrical capacity of the motor of each piece of equipment. The values contained in the information on amperes and volts were used for the calculation, because according to Ueno the formula for calculating power (W) is: P=Ui, where U represents volts and i represents ampere values (2006).

To transform watts (W) into kilowatts (kW), you need to take the value of (W) and divide by 1000, thus obtaining the KW (UENO, 2006). To calculate the cost of energy, we used the value of the energy bill, model appendix D, to calculate the average cost of each KW spent during the month. The unit cost of the KW is multiplied by the total quantity of KW/h used in the month to obtain the monthly cost of the equipment.

To calculate the KW/h you need to calculate the kilowatt and multiply it by the number of hours used in the month (CARRON, GUIMARÂES, 2003, p.235).

Data in ampere and volt values for each piece of equipment:

Equipment	Amps, Watts	Volts	Kw/h
Milking machine	8,5 A	220	1,87
Cooler	4,4 A	220	0,968
Fodder chopper	12,5 A	220	82,5
Grain crusher	12,5 A	220	82,5
Lamp	60 W	220	0,06

APPENDIX D - COST OF SALT CONSUMPTION

Common salt cost			
Mes	**Purchase price**	**Period months**	**Monthly Value**
May	R$ 13,95	3	R$ 4,65
August	R$ 13,00	4	R$ 3,25
December	R$ 13,00	3	R$ 4,33
Margo	R$ 15,00	3	R$ 5,00

Mineral Salt Cost			
Mes	**Value**	**Period months**	**Monthly Value**
Apr/12	R$ 20,00	8	R$ 2.50ze
Dec/12	R$ 25,56	7	R$ 3,65

Monthly Salary Cost			
Mes	**Value**	**Period months**	**Monthly Value**
May	R$22,52	3	R$ 7,51

August	R$ 24,43	4	R$ 6,11
December	R$ 23,95	3	R$ 7,98
Margo	R$ 25,95	3	R$ 8,65

APPENDIX E - COST OF QUIRERA

Corn planting costs		
Inputs	**Quantity**	**Total**
Fertilizer	5	R$315,00
Seed	1	R$ 70,00
Tractor driver	1	R$ 70,00
Poison	1	R$ 50,00
Urea	1,5	R$ 93,00
Total		R$598,00
Harvester	1	R$180,00
Transport. Corn	1	R$25,00
Canvas	1	R$ 80,00

total harvest cost		R$285,00
Total Cost	50 bags	R$883,00
Monthly cost per bag	60 Kg	R$17,66

Monthly cost			
Month	Quantity of bags	Bag cost	Monthly cost
May/12	3,5	R$ 17,66	R$ 61,81
Jun/12	3,5	R$ 17,66	R$ 61,81
Jul/12	3,5	R$ 17,66	R$ 61,81
Aug/12	3,5	R$ 17,66	R$ 61,81
Sep/12	3,5	R$ 17,66	R$ 61,81
Oct/12	3,5	R$ 17,66	R$ 61,81
Nov/12	3,5	R$ 17,66	R$ 61,81
Dec/12	3,5	R$ 17,66	R$ 61,81

Jan/13	3,5	R$ 17,66	R$ 61,81
Feb/13	3,5	R$ 17,66	R$ 61,81
Mar/13	3,5	R$ 17,66	R$ 61,81
Apr/13	3,5	R$ 17,66	R$ 61,81

APPENDIX F - COST OF MILKING OIL

Milking machine oil cost			
Month	**Value**	**Period months**	**Monthly cost**
Apr./12	R$ 9,66	3	R$ 3,22
Jul./12	R$ 9,66	2	R$ 4,83
Sep./12	R$ 9,81	3	R$ 3,27
Dec./12	R$ 9,81	2	R$ 4,91
feb./13	R$10,00	2	R$ 5,00
Apr./13	R$10,00	2	R$ 5,00

APPENDIX G - DEPRECIATION CALCULATION

Depreciating	Milking machine	Cooler	Structures	Safeguarding	Fodder chopper	Shredder/grinder	Matrices
Good value	R$ 1.200,00	R$ 500,00	R$ 5.000,00	R$ 1.500,00	R$150 ,00	R$800 ,00	R$17 .000,00
Years of use	9	11	19	19	7	11	
Useful life	10	10	25	25	10	12	10
Residual value	-	-	20%	20%	5%	5%	40%
Depreciated year	10%	10%	4%	4%	10%	8,33%	10%
Depreciated month	0,83%	0,83%	0,33%	0,33%	0,83%	0,69%	0,83%
Annual depreciation value	R$ 120,00	-	R$ 160,00	R$ 48,00	R$14 ,20	R$63 ,33	R$1 .020,00
Monthly depreciation value	R$ 10,00	-	R$ 13,33	R$ 4,00	R$1 ,18	R$5 ,28	R$85 ,00

Calculation Depreciating Values from May 2012 to April 2013

Depreciation Matrices

Matrices	Good value	Years of use	Useful life	residual value	Annual depreciation	deprec. value Annual	deprec. value Monthly
1	R$ 2.000,00	7	10	40%	10%	R$ 120,00	R$ 10,00
2	R$ 2.500,00	7	10	40%	10%	R$ 150,00	R$ 12,50
3	R$ 2.500,00	6	10	40%	10%	R$ 150,00	R$ 12,50
4	R$ 2.000,00	6	10	40%	10%	R$ 120,00	R$ 10,00
5	R$ 2.000,00	5	10	40%	10%	R$ 120,00	R$ 10,00
6	R$ 2.000,00	4	10	40%	10%	R$ 120,00	R$ 10,00
7	R$ 2.000,00	4	10	40%	10%	R$ 120,00	R$ 10,00
8	R$ 2.000,00	4	10	40%	10%	R$ 120,00	R$ 10,00
Total depreciated	R$ 17.000,00		10	40%	10%	R$ 1.020,00	R$ 85,00

APPENDIX H - COST OF SOYBEAN PRODUCTION

COST OF PRODUCTION SOYBEAN CROP 2011/2012			
Inputs	**Quantity**	**Unit Value**	**Total**
Land preparation			**R$ 337,71**
Drying plant tractor	1	R$ 100,00	R$ 100,00
Herbicide gramoxo	5.	R$15,00	R$75,00
Staron herbicide	2	R$18,00	R$36,00
Roundap herbicide	5	R$13,00	R$65,00
Roundap	1	R$61,71	R$61,71
Plantation			**R$1.451,48**

Fertilizer	15	R$53,00	R$ 795,00
Cooper soybean seed.	4	R$61,12	R$ 244,48
Dinon soybean seed	1	R$72,00	R$72,00
Seed treatment	1	R$60,00	R$60,00
Tractor Planting	4	R$70,00	R$ 280,00
Plant treatment			**R$ 392,94**
Priori xtra Rust	1	R$ 118,69	R$ 118,69
Curion	1	R$67,84	R$67,84
Green grain - ant	1	R$8,00	R$8,00
Roundap herbicide - weeding	5	R$12,00	R$60,00
Spreading oil	1	R$38,41	R$38,41

Tractor driver	1	R$ 100,00	R$ 100,00
Harvesting	1	12%	14,4
Harvest costs	14,4	R$50,00	**R$ 720,00**
Total			**R$2.902,13**

Unit cost				
Safra	**Number of bags harvested**	**Quantity KG**	**Total cost**	**Unit cost KG**
2011/2012	120	7200	R$2.902,13	R$0,40

COST OF PRODUCTION SOYBEAN CROP 2012/2013			
Inputs	**Quantity**	**Unit Value**	**Total**
Land preparation			**R$ 191,00**

Drying plant tractor	1	R$ 110,00	R$ 110,00
Roundap herbicide	5	R$ 15,00	R$75,00
Transportation of inputs	4	R$1,50	R$6,00
Plantation			**R$1.953,00**
Fertilizer	19	R$56,00	R$1.064,00
Fertilizer	1	R$64,00	R$64,00
Seed	5	R$ 105,00	R$ 525,00
Tractor driver	1	R$ 300,00	R$ 300,00
Plant care			**R$ 649,63**
Roundap herbicide	5	R$17,00	R$85,00

Tractor driver	1	R$ 100,00	R$ 100,00
Priori xtra rust	2	R$ 131,81	R$ 263,62
Talstar caterpillar and fede-fede	1	R$77,42	R$77,42
Green grain - ant	1	R$3,59	R$3,59
Tractor driver	1	R$ 120,00	R$ 120,00
Harvesting	1	12%	20,4
Harvest value	20,4	R$56,00	R$1.142,40
Total			**R$3.936,03**

Unit cost				
Safra	**Number of bags harvested**	**Quantity KG**	**Total cost**	**Unit cost KG**
2012/2013	175	10500	R$3936,03	R$0,37

ANNEXES
ANNEX A - DEPRECIATION TABLES

Depreciated Table (	Adapted from Conab 2010, p.52-57)	
Specified	**Useful life (years)**	**Residual value**
Wooden shed	25	20%
Fence	25	20%
Fodder chopper	10	5%
Crusher, grinder and chopper	12	5%

Depreciation Table - (Adapted from MARION 2010, p.52-55)		
Specification	**Useful life (years)**	**Depreciation rate**
Milking machine	10	10%

Depreciation Table (Adapted from SRF No. 162)			
Code	**Specified**	**Useful life**	**Annual depreciation**

| 8418 | REFRIGERATORS, FREEZERS AND OTHER REFRIGERATING OR FREEZING EQUIPMENT, ELECTRIC OR OTHER; HEAT PUMPS OTHER THAN AIR CONDITIONING MACHINES OF HEADING 8415 | 10 | 10% |

ANNEX B - INVOICE FOR THE SALE OF MILK FROM PRODUCTION IN DECEMBER/2012

LATICINIOS SANTO CRISTO LTDA

RST 472, KM01 - CONTORNO SANTO CRISTO, n° 0, 98960000, INTERIOR, SANTO CRISTO, RS

DANFE — DOCUMENTO AUXILIAR DA NOTA FISCAL ELETRÔNICA

0 - ENTRADA
1 - SAÍDA   0

FOLHA 1 DE 1

CHAVE DE ACESSO
4313 0105 2482 4200 0112 5500 0000 0184 0412 6892 2650

Consulta de autenticidade no portal nacional da NF-e www.nfe.fazenda.gov.br/portal ou no site da Sefaz autorizadora

NATUREZA DA OPERAÇÃO: CONTRA NOTA DE PRODUTOR

INSCRIÇÃO ESTADUAL: 1160017180

PROTOCOLO DE AUTORIZAÇÃO DE USO: 143130005600709   14/01/2013 10:47:13

CNPJ: 05248242000112

DESTINATÁRIO / REMETENTE

NOME / RAZÃO SOCIAL: EDEGAR EBERHARDT EOU ODETE SIGOLIN | 1333
CPF/CNPJ: 33228680010
DATA DE EMISSÃO: 14/01/2013

ENDEREÇO: L 17 SETEMBRO NORTE, 0
BAIRRO / DISTRITO: INTERIOR
CEP: 93900000
DATA DE SAÍDA/ENTRADA:

MUNICÍPIO: SANTA ROSA
UF: RS
INSCRIÇÃO ESTADUAL: 1101057707
HORA DE SAÍDA:

FATURA / DUPLICATAS

FORMA DE PAGAMENTO: Pagamento a Vista
NÚMERO DA FATURA: 0
VALOR ORIGINAL: 1 522,63
DESCONTO: 0,00
VALOR LÍQUIDO: 1 522,63

CÁLCULO DE IMPOSTO

VALOR TOTAL DOS PRODUTOS: 1 559,40
VALOR TOTAL DA NF-e: 1 522,63

TRANSPORTADOR / VOLUMES TRANSPORTADOS

RAZÃO SOCIAL: TRANSPORTES IGO LTDA
FRETE POR CONTA: 0
PLACA DO VEÍCULO: IFQ8000
UF: RS
CNPJ/CPF: 05556958000187

ENDEREÇO: RUA JOSE ALFREDO NEDERL 1175
MUNICÍPIO: SANTA ROSA
UF: RS
INSCRIÇÃO ESTADUAL: 1100091200

QUANTIDADE: 0
ESPÉCIE: VOLUMES
PESO BRUTO: 0,00
PESO LÍQUIDO: 0,00

DADOS DO PRODUTOS/SERVIÇOS

CÁLCULO DO ISSQN

VALOR TOTAL DOS SERVIÇOS: 0,00
BASE DE CÁLCULO DO ISSQN: 0,00
VALOR DO ISSQN: 0,00

DADOS ADICIONAIS

RESERVADO AO FISCO

Nota Fiscal Eletrônica disponível para download em: http://nfe.abasesul.com.br/nfeGestor

RECEBEMOS DE LATICINIOS SANTO CRISTO LTDA OS PRODUTOS/SERVIÇOS CONSTANTES DA NF-e INDICADA AO LADO, EMITIDA EM 14/01/2013

EDEGAR EBERHARDT EOU ODETE SIGOLIN | 1333

VALOR TOTAL DA NF-e: R$ 1.522,63

NF-e
N°: 000.018.404
SÉRIE: 000

DATA DE RECEBIMENTO | IDENTIFICAÇÃO E ASSINATURA DO RECEBEDOR

TRANSPORTES IGO LTDA

PARCERIA

TCHÊ MILK

VILA 7 DE SETEMBRO - SANTA ROSA - F.: 3505-8116

CONTROLE DE PESAGEM DE LEITE

Fornecedor: _______________________________ Nº: 08

End.: ____

MÊS: DEZEMBRO

Dia	Litros	Dia	Litros
01		17	
02	158	18	154
03		19	
04	158	20	166
05		21	
06	162	22	135
07		23	
08	150	24	153
09		25	
10	145	26	147
11		27	
12	148	28	142
13		29	
14	152	30	140
15		31	
16	150	TOTAL DO MÊS	2260

ANNEX D - PROPERTY ELECTRICITY BILL

COOPERLUZ - COOPERATIVA DISTRIBUIDORA DE ENERGIA FRONTEIRA NOROESTE
MATRIZ RUA BELA VISTA, N° 52 • SANTA ROSA • RS
SEDE ADM. AV. SANTA CRUZ N° 989 • FONE: (55) 3511-9500
CNPJ: 95.604.760/0001-41 • INSCRIÇÃO ESTADUAL: 1190017515

NOTA FISCAL CONTA DE ENERGIA ELÉTRICA
SÉRIE ÚNICA

N.º 000 007.591

N° DE CONTROLE DO FORMULÁRIO
N° 516103

SEU NÚMERO CONOSCO → **1235101**

CFOP: 5 256

UNIDADE CONSUMIDORA

EDGAR EBERHARDT
Matrícula: 12351   Ligação: 01
LINHA 7 DE SETEMBRO NORTE, S/N

INTERIOR
98900-000 SANTA ROSA - RS
CNPJ/CPF: 332.286.800-10
Inscr Est./CI:1017911627
Classe: RURAL

Tarifa: B2
N° Prod: 1101057707
Subclasse: Agropecuária Rural

DADOS DE FATURAMENTO

Emissão 28/03/2013	Apresentação	01/04/2013
Mês/Ano Faturamento		03/2013
Leitura Atual	12/03/2013	61.417
Leitura Anterior	08/02/2013	60.986
Consumo Faturado (kWh)	X 1.000 =	431
Consumo Diário (kWh)		13,47
Dias de Consumo		32
Ocorrência do mês		Lido
Próxima Leitura		09/04/2013
Valor TUSD:	69,69 Valor TE:	58,87

DADOS TÉCNICOS E COMERCIAIS DE MEDIÇÃO

Instalação transformadora	10566
Número do Medidor	9208913
Fator de Multiplicação	1.000
Tipo de Ligação	Monofásico
Kva disponível	17.7

Tensão de fornecimento: Baixa 220 V - Monof. V Trifásica / Limites de tensão adequada Mínima - 201 V Máxima - 231 V

ITENS FATURADOS

Consumo de energia	Qtde.	Tarifa (R$/kWh)	Valor R$
Energia Ativa	100	0.33897	33.89
Energia Ativa	331	0.29829	98.73
Subtotal			132.62

HISTÓRICO DE CONSUMO

Mês/ano	Dias	Ocorrência	Pagto.	Valor	Leitura	kWh
03/2013	32	Lido	Aberto	132.62	61417	431
02/2013	29	Lido	03/2013	113.83	60986	366
01/2013	33	Lido	02/2013	141.27	60618	460
12/2012	29	Lido	01/2013	126.49	60158	402
11/2012	31	Lido	12/2012	115.39	59756	365
10/2012	31	Lido	11/2012	102.31	59391	325
09/2012	32	Lido	10/2012	115.82	59066	369
08/2012	28	Lido	09/2012	95.33	58697	300
07/2012	29	Lido	08/2012	96.00	58397	310
06/2012	33	Lido	07/2012	116.13	58087	383
05/2012	29	Lido	06/2012	90.31	57704	328
04/2012	33	Lido	05/2012	133.60	57376	445
03/2012	30	Lido	04/2012	130.22	56931	438

INDICADORES DE CONTINUIDADE DO FORNECIMENTO

Conjunto: 1 Cooperluz - Santa Rosa   Competência: 02/2013

	DIC	FIC	DMIC	DEC	FEC
METAS	12.76	8.04	6.99	7.6	4.50
REALIZADOS	1.83	2	52	4.64	3.27

Componentes de Custo de Energia R$:

Distribuição	98.99	ICMS	4.06
Transmissão	2.96	PIS	0.00
Energia	12.21	COFINS	0.00
Encargos	14.40	Outros	0.00

VENCIMENTO	TOTAL A PAGAR
25/04/2013	**132,62**

FATURAS PENDENTES

ICMS INCLUSO NO TOTAL DA FATURA

Base R$	33.8
Aliq. do ICMS	12
VALOR R$	4.0

Atraso de pagamento será cobrado multa de 2% + juros de mora, a serem cobrados em conta posterior. Sujeito a suspensão do fornecimento após 15 dias do vencimento.

RESERVADO AO FISCO

B084.BBD6.5DE5.0D19.93E0.DEA9.F677.6B10

ANNEX E - WHEAT PRICE QUOTATION FOR OCTOBER 2012.

Wheat PH 78 Quota

Date	Price in R$	
31/10/2012	R$	31,00
30/10/2012	R$	31,00
29/10/2012	R$	31,00
26/10/2012	R$	32,00
25/10/2012	R$	32,00
24/10/2012	R$	32,00
23/10/2012	R$	32,00
22/10/2012	R$	32,00
19/10/2012	R$	32,00
18/10/2012	R$	32,00
17/10/2012	R$	31,00
15/10/2012	R$	31,00
11/10/2012	R$	31,00
10/10/2012	R$	31,00
14/09/2012	R$	27,00

The wheat price for October 2012 is sourced from the Cotrirosa Triticola Santa Rosa website.

ANNEX F- PRICE QUOTATION FOR A SACK OF SOYBEANS FOR THE MONTH OF

JUNE 2013

Soy	
Date	Price in R$
14/06/2013	60,00
13/06/2013	61,00
12/06/2013	61,00
11/06/2013	60,00
10/06/2013	60,00

07/06/2013	60,00
06/06/2013	59,50
05/06/2013	59,50
04/06/2013	59,50
03/06/2013	59,00
31/05/2013	58,00
29/05/2013	57,00
28/05/2013	56,00
27/05/2013	56,00
24/05/2013	57,00

Soybean bag nail sold at the Cotrirosa Cooperative.
Information based on Soy quotations, from Cotrirosa's email address.

ANNEX G - PRICE OF A SACK OF SOYBEANS IN APRIL 2012

Soy	
Date	Price in R$
23/04/2012	52,50
20/04/2012	52,00
19/04/2012	51,50
18/04/2012	51,50
17/04/2012	51,50
16/04/2012	51,50

Date	Price in R$
13/04/2012	51,50
12/04/2012	51,00
11/04/2012	51,00
10/04/2012	51,00
09/04/2012	51,00
05/04/2012	51,00
04/04/2012	51,00
03/04/2012	51,00
02/04/2012	50,50

Soybean bag nail sold at the Cotrirosa Cooperative.
Information based on Soy quotations, from Cotrirosa's email address.

ANNEX H - PRICE OF A SACK OF SOYBEANS FOR THE MONTH OF APRIL 2013.

Soy		
Date	Price in R$	
11/04/2013	R$	53,00
10/04/2013	R$	53,00
09/04/2013	R$	52,00
08/04/2013	R$	52,00
05/04/2013	R$	52,00
04/04/2013	R$	53,00
03/04/2013	R$	54,00

02/04/2013	R$	54,00
01/04/2013	R$	54,00
28/03/2013	R$	55,00
27/03/2013	R$	54,00
26/03/2013	R$	54,00
25/03/2013	R$	54,00
22/03/2013	R$	54,00
21/03/2013	R$	54,00

Soybean bag nail sold at the Cotrirosa Cooperative.
Information based on Soy quotations, from Cotrirosa's e-mail address.